NEW TOP HOTEL

顶级新酒店　　深圳市创扬文化传播有限公司 / 编　　卢晓娟 张秋 和精伟 刘洋 郑希彬 / 译

大连理工大学出版社
Dalian University of Technology Press

图书在版编目 (CIP) 数据

顶级新酒店 / 深圳市创扬文化传播有限公司编；卢晓娟等译. —大连：大连理工大学出版社, 2013.7
ISBN 978-7-5611-7856-0

Ⅰ. ①顶… Ⅱ. ①深… ②卢… Ⅲ. ①旅游饭店 – 室内装饰设计 – 世界 – 图集 Ⅳ. ① TU247.4

中国版本图书馆 CIP 数据核字 (2013) 第 105202 号

出版发行：大连理工大学出版社
　　　　　（地址：大连市软件园路 80 号 邮编：116023）
印　　刷：上海锦良印刷厂
幅面尺寸：240mm × 320mm
印　　张：20
插　　页：4
出版时间：2013 年 7 月第 1 版
印刷时间：2013 年 7 月第 1 次印刷
策划编辑：袁　斌　刘　蓉
责任编辑：刘　蓉
责任校对：王丹丹
封面设计：四季设计

ISBN 978-7-5611-7856-0
定　　价：328.00 元

电话：0411-84708842
传真：0411-84701466
邮购：0411-84703636
E-mail:designbooks_dutp@yahoo.com.cn
URL:http://www.dutp.cn

如有质量问题请联系出版中心：（0411）84709246　84709043

NEW TOP HOTEL

CONTENTS
目 录

006-017	DREAM DOWNTOWN HOTEL	梦幻市区酒店
018-031	HILTON PATTAYA HOTEL	希尔顿芭提雅酒店
032-049	HORIZON HOTEL	地平线酒店
050-057	RICA HOTEL NARVIK	纳尔维克利卡酒店
058-067	W BALI VILLAS AND E-WOW SUITE	巴厘岛W酒店别墅和E-WOW套房

068-075	W HOTEL – SAN DIEGO	圣地亚哥W酒店
076-087	ONSEN PAPAWAQA	泰安观止
088-093	T HOTEL	T酒店
094-103	TOWN@HOUSE STREET HOTEL	Town@House街区旅馆
104-107	TOWN HOUSE STREET MILANO DUOMO	市政街米兰迎宾馆

108-117	BLOSSOM HILL – ZHOUZHUANG SEASONLAND	花间堂 · 周庄季香院
118-125	GRAND HYATT SHENZHEN HOTEL	深圳君悦酒店
126-137	ONE & ONLY REETHI RAH	One & Only瑞提娜岛度假村
138-143	NEVAI HOTEL	那威酒店
144-151	HOTEL MISSONI KUWAIT	科威特米索尼酒店
152-159	RADISSON BLU AQUA	雷迪森丽笙水绿酒店
160-165	NINGBO RITZ HOTEL	宁波丽兹酒店
166-171	HOTEL NIKKO SHANGHAI	上海日航饭店

172-185	BEIJING SHOUZHOU HOTEL	北京寿州大饭店
186-195	LI MING HOTEL	黎明大酒店
196-201	25HOURS HOTEL IN VIENNA	维也纳25小时酒店
202-207	CONSERVATORIUM HOTEL	音乐学院酒店
208-213	KAMEHA GRAND BONN	波恩卡梅哈大酒店
214-223	KEMPINSKI HOTEL	凯宾斯基酒店
224-229	B.O.G. HOTEL	B.O.G.酒店

230-237	SALA PHUKET RESORT AND SPA	普吉岛莎拉度假村
238-245	THE HOUSE HOTEL BOSPHORUS	博斯普鲁斯House酒店
246-253	DESIGN BOUTIQUE HOTEL	蝶尚非经验酒店
254-263	RAMADA PLAZA OPTICS VALLEY HOTEL WUHAN	武汉华美达光谷大酒店
264-271	SHANGRI-LA'S FAR EASTERN PLAZA HOTEL, TAINAN	香格里拉台南远东国际大饭店

272-279	ICELANDAIR HOTEL REYKJAVIK MARINA	雷克雅未克码头冰岛航空酒店
280-287	UNA HOTEL	尤纳酒店
288-297	NANJING 21° THEME HOTEL	南京21°主题酒店
298-305	JIMBARAN HOTEL IN XIAMEN	厦门金巴兰酒店
306-319	W HOTEL, LEICESTER SQUARE, LONDON	伦敦莱斯特广场W酒店

DREAM DOWNTOWN HOTEL 梦幻市区酒店

Architect/
Handel Architects - Frank Fusaro, AIA, Partner
Location/
New York City, USA
Client/
Hampshire Hotels & Resorts + Vikram Chatwal Hotels
Site Area/
2,350m²
Gross Floor Area/
17,159m²
Photographer/
Bruce Damonte, Philip Ennis

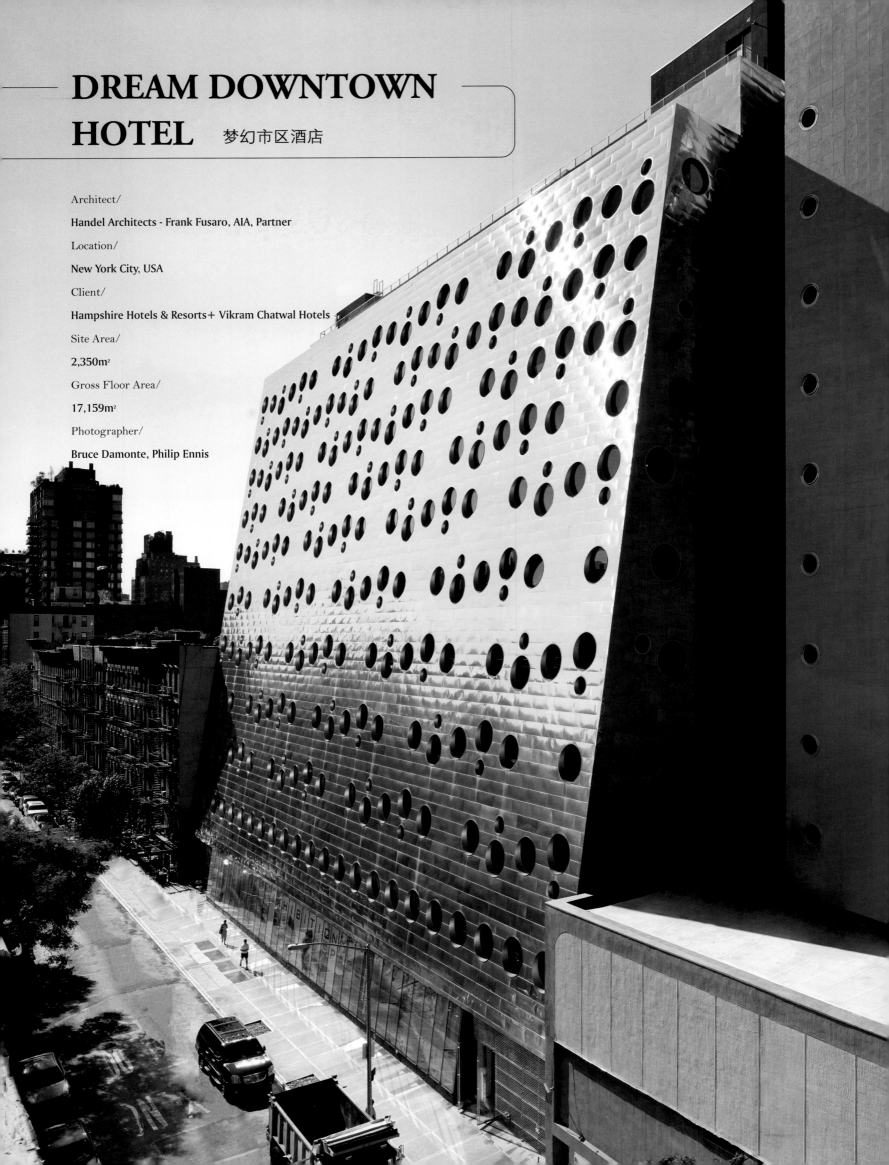

Dream Downtown Hotel is a 17,159m² boutique hotel in the Chelsea neighborhood of New York City. The 12-story building includes 316 guestrooms, two restaurants, rooftop and VIP lounges, outdoor pool and pool bar, a gym, event space, and ground floor retail.

Dream sits on a through-block site, fronting both 16th and 17th Streets, and is adjacent to the Maritime Hotel, which sits adjacent to the west. The otherness of Ledner's 1966 design for the National Maritime Annex was critical to preserve. New porthole windows were added, loosening the rigid grid of the previous design, while creating a new facade of controlled chaos and verve. The tiles reflect the sky, the sun, the and moon, and when the light hits the facade perfectly, the stainless steel disintegrates and the circular windows appear to float like bubbles. The orthogonal panels fold at the corners, continuing the slope and generating a contrasting effect to the window pattern of the north facade.

The original through-block building offered limited possibilities for natural light. Four floors were removed from the center of the building, which created a new pool terrace and beach along with new windows and balconies for guestrooms. The glass bottom pool allows guests in the lobby glimpse through the water to the outside (and vice versa) connecting the spaces in an ethereal way. Light wells framed in teak between the lobby, pool and lower levels allow the space to flow. Two hundred hand blown glass globes float through the lobby and congregate over The Marble Lane restaurant filling the space with a magical light cloud. Fixtures and furnishings were custom-designed for the public spaces and guestrooms to complement the exterior design and to continue the limitless feeling of space throughout the guest experience.

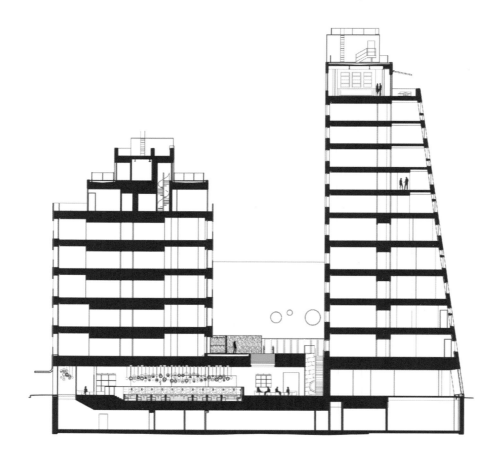

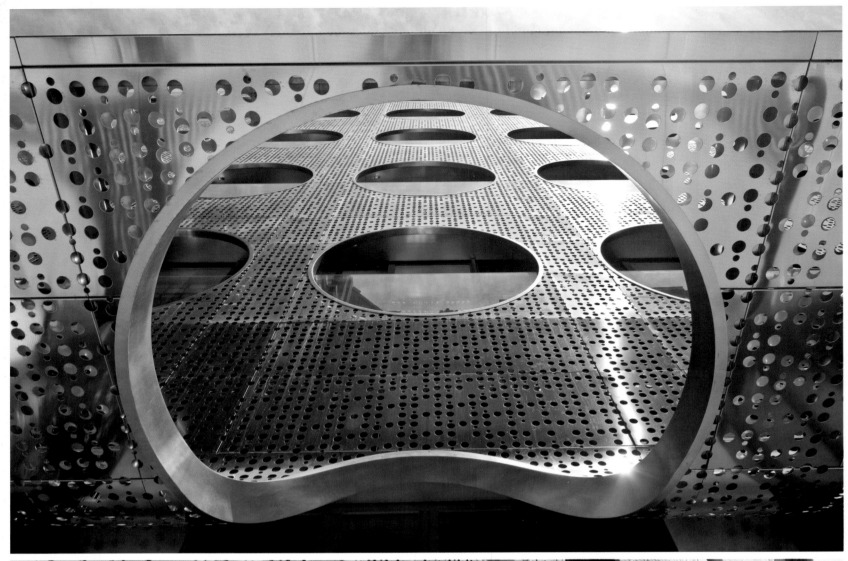

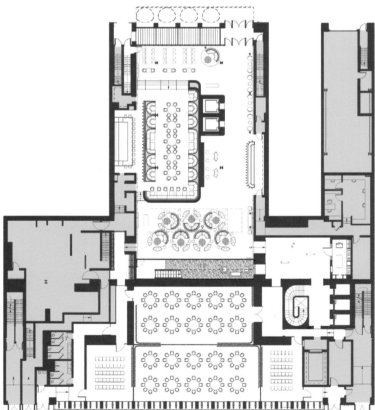

梦幻市区酒店是一家建筑面积为17159平方米的精品酒店，位于纽约切尔西附近。酒店共12层，包括316间客房、两间餐厅、屋顶花园和贵宾休息室、室外泳池和池边酒吧、健身房、活动大厅以及一层的零售商业区。

梦幻市区酒店横跨街区而建，面向16号街和17号街，毗邻靠近西部的滨海酒店。这个由莱德纳于1966年为国家海事协会设计的另一座附属建筑急待保护。酒店新增加了舷窗式窗户，缓解了原来设计中格子窗的死板感。新的酒店外立面看似无序却是可控的，充满了活力。墙砖映射出天空以及日月的光华，当阳光完全照射到酒店外立面时，不锈钢外表呈现出破碎状，圆形的窗户就像气泡般浮动。矩形嵌板在拐角处折叠，继续保持坡度，与酒店北立面的窗户格局产生对比效果。

原有的跨街区建筑采光效果有限。从建筑的中心减去四层，腾出空间建造了一个新的泳池平台和憩息沙滩以及新开的窗子和客房阳台。玻璃底泳池让大厅的宾客能透过池水看到外面（反过来也一样），让空间的连接有一种虚无缥缈的感觉。在大厅、泳池和低楼层间的柚木框架采光井使空间具有流动感。200个手工吹制的玻璃球体飘浮在大厅中，聚集在理石巷餐厅上空，像魔幻光云充斥着整个空间。酒店设施和用品为公共空间和客房设计定制，以完善酒店的外观设计，不断为客人提供无限的空间体验。

HILTON PATTAYA HOTEL 希尔顿芭提雅酒店

Design Firm/
Department of ARCHITECTURE Co., Ltd.

Location/
Chonburi, Thailand

Photographer/
Wison Tungthunya

Lobby & "DRIFT" Bar

The architectural intervention to the entire ceiling plane, with its dynamic wave lines, leads the movement of the arriving visitors towards the seafront beyond. The fabric installation on the ceiling becomes a main feature in the space, when at night, strip lighting accents from the fabric linear pattern above. The whole ceiling volume becomes a gentle luminous source of light giving a fine ambient to the overall space.

At the end of the lobby space, the bar area is arranged linearly along the building edge parallel to the sea with maximum opening to the ocean view. Backdrop of the bar area lies a wooden wall with alcoves where the daybeds partially tuck themselves into the wall.

Further in front of the indoor bar area is an outdoor lounge space with a large reflecting pond catching the reflection of both the sky and the droplet daybeds and lamps scattered around.

"FLARE" Fine Dining

The design explores a mediating means of space demarcation between pockets of private dining area. By occupying an intermediary space between spaces with a translucent volume of sheer fabric, the effect results in an elegant, mystifying atmosphere, engaging and disengaging different spaces at the same time. A slight up-light to the volume of sheer fabric and the glowing light at the edge of the fabric accentuated the lighting in the space.

大堂和"Drift"酒吧

对整个天花板部分的建筑学处理,搭配动态的波浪线条,将入住旅客的目光引向远处的海滨。天花板上的装饰布是酒店的主要特色。夜晚来临,条形的灯光从线形装饰布中凸显出来。整个天花板成了柔和的光源,为整个空间营造出美好的氛围。

在大堂的尽头,酒吧区域沿着建筑物边缘呈直线排列,与海岸平行。在这里,客人可以最大限度地欣赏大海的景色。酒吧的背景墙是带有凹室的木制墙壁,坐卧两用的长沙发有部分设计在凹室内。

室内酒吧区再往前是户外休闲区域,那里有一个镜面般的大水池,池里倒映着天空、水滴状的坐卧两用沙发以及摆放在各处的地灯。

"Flare"高级餐厅

这个设计探索了一种将整个空间分隔成小块私人就餐区的方法。设计师用半透明的薄纱将各个区域隔开,满足不同空间在同一时间自由利用的需求,营造出高雅又神秘的氛围。照射到薄纱上的微弱灯光和薄纱边缘的光线加强了空间的照明。

HORIZON HOTEL

地平线酒店

Designer/

Botta lai/SCALE

Location/

K.K Sabah Island, Malaysia

Area/

15,895m²

Photographer/

Deep Blue Photo Club

Horizon Hotel is located in K.K Sabah Island, Malaysia, occupied the beautiful coastline. It consists of Chinese and Western restaurants, coffee hall, fitness clubs and SPA seaview swimming pool. Guests can enjoy both top food and a panoramic view of Sabah Island. From the presidential suite to different types of rooms and meeting halls, the functionality is maximized to meet the different needs of guests.

In the case of introducing "Cabinet construction" box construction theory, the interior space is permeated into each other, to form "Borrowed scenery" and visual effect. 1F–2F space bridges with lighting design and the boxes suspended in the air, allows the LOBBY to get full light, at the same time to achieve the visual impact. Guest room corridors, exaggerated black piano baking varnish Deco door head, with the wave totem of the carpet, and works from local photographer are used to decor the space. Designers use contemporary design, integrating into the local art, to enhance the intrinsic culture of Horizon Hotel.

Horizon Hotel has 9 types of rooms, with a total of 183, in which Deco suite is the highlighting point. The designer specializes in the use of local photographers' works for the space to elaborate continuation of the space, hand sewn soft bag, ebony floor, with a yuppie style stool, to reflect both the low-key luxury and cultural basis. SPA room includes a spa, a steam room, a sauna room and a massage room, with perfect functions, allowing guests to relax and enjoy a moment of peace.

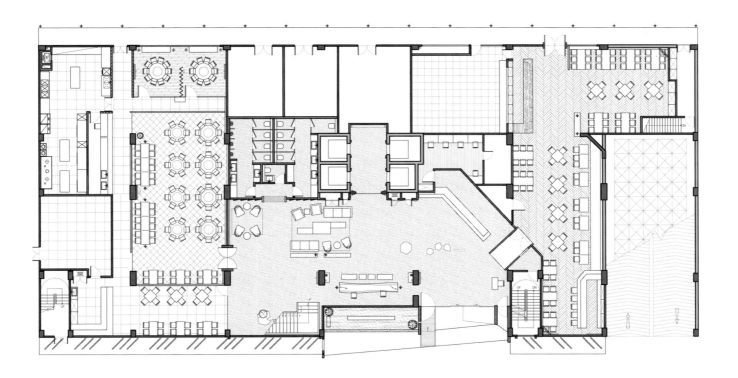

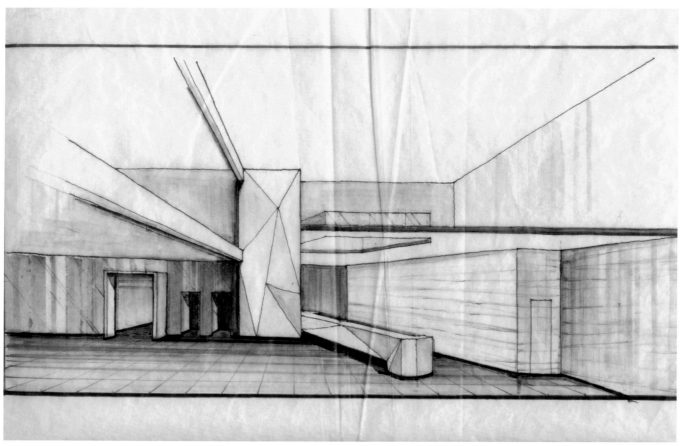

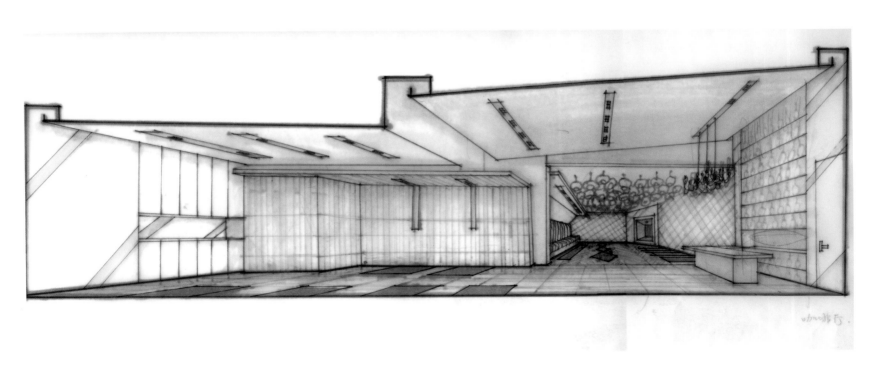

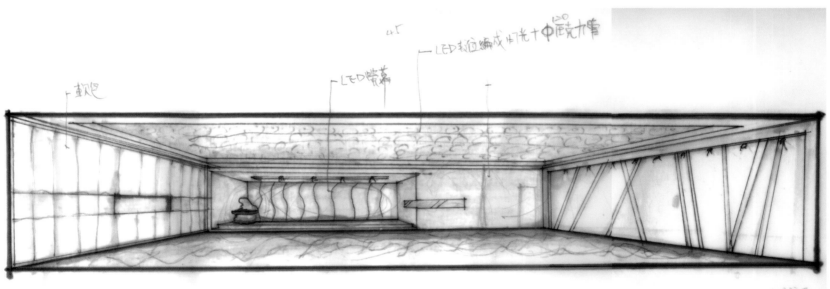

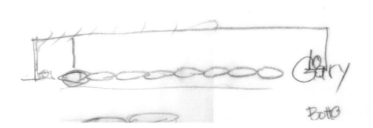

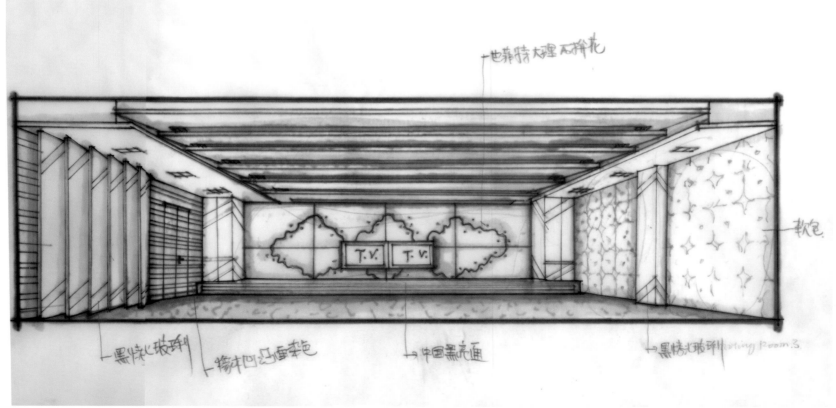

　　地平线酒店坐落于马来西亚K.K沙巴岛，占据唯美海岸线。酒店拥有中西餐厅、咖啡厅、健身会所及SPA全海景游泳池。宾客享用顶级美食之余，更可将沙巴岛景观尽收眼底。从总统套房到不同户型客房和多规格的会议厅让酒店的实用功能最大化，可满足宾客的不同需求。

　　本案引入了Cabinet Construction箱体建筑理论，让室内空间相互渗透，在视觉上形成"借景"、"对视"的效果。一层至二层的空间廊桥配合灯光设计和悬浮于空中的箱体，给予大厅充分光线的同时，达到视觉冲击的效果。客房廊道、夸张的黑色钢琴烤漆Deco门头，搭配海浪图腾的地毯，并取当地摄影家作品为素材装饰空间。设计师用当代设计手法，融入当地艺术，以提升精品酒店的内在文化。

　　地平线酒店拥有9种房型，共183间，其中Deco房型为套房中的亮点。设计师擅长运用当地摄影师作品为空间做铺陈，延续空间景深；手工车缝软包、黑檀木地板，搭配雅痞风格矮凳，体现低调奢华的同时又展现蕴藏于本案的文化底蕴。SPA房间包括温泉室、蒸汽浴室、桑拿浴室和按摩室，功能完善，让宾客放松心境，独享片刻安宁。

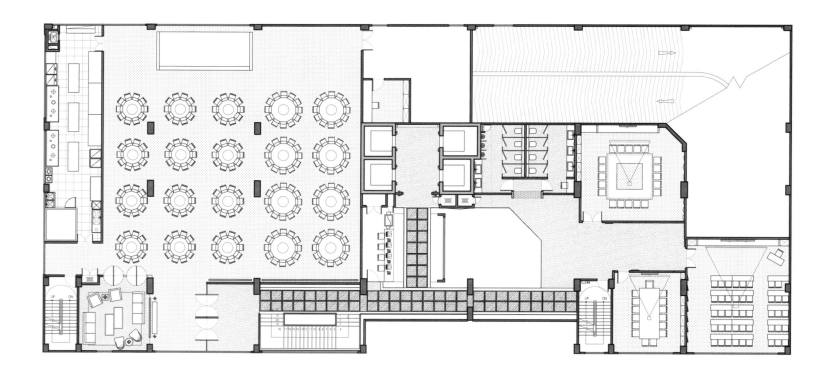

RICA HOTEL NARVIK 纳尔维克利卡酒店

Interior Design Office/

AS Scenario Interiørarkitekter MNIL

Lead Designer/

Annethe Thorsrud Interior Architect BA Hons

Location/

Kongensgate 33 Narvik, 8500, Norway

Area/

6,220m²

Photographer/

Gatis Rozenfelds, F64 SIA (Latvia)

Rica Hotel Narvik, Northern Norway's highest building.

The architectural form that towers over 18 floors gives this hotel a wonderful view of the city, the fjord "Ofotfjorden", the Mountain "Dronningen" and the other surrounding mountains. Rica Hotel Narvik has a stylish interior that takes up the exciting exterior of the 148 luxurious and spacious guest rooms, all-day restaurant, and one floor sky bar with panoramic terrace on the top floor. The hotel arranges for business travelers with eight functional meeting rooms, where the largest room has a capacity for up to 175 people.

The interior has a modern look that plays on the natural elements found in Narvik. Inspiration is drawn from nature, where you will find neutral base color and bright color accents and wood. Along with large graphic patterns in vibrant color combinations in both carpets, wall patterns against the wooden look, the designers "break" up the tight style of the exterior of the building, to create a calm, warm and colorful atmosphere inside. The furniture is mainly modern and clean with bright color combinations. Comfort, durability and of course the design were essential elements for the selection of the furniture. The Client chose to work with local artists for art & decor, for both rooms and common areas.

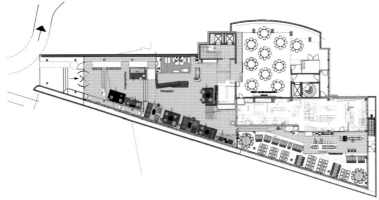

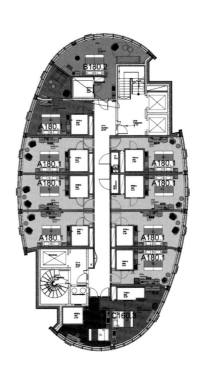

　　纳尔维克利卡酒店是挪威北部最高的建筑。

　　超过18层楼的建筑形式使这座酒店拥有得天独厚的观景条件。Ofotfjorden峡湾、Dronningen山以及其他山脉都尽收眼底。纳尔维克利卡酒店内部设计时尚，表面设计大胆创新，拥有148间奢华宽敞的客房、全天营业的餐厅以及顶层一整层的空中酒吧兼全景观赏台。该酒店还为商务旅行者配备了8间功能性会议室，其中最大的会议室至多可容纳175人。

　　酒店内部现代的设计融合了纳尔维克的自然元素。设计灵感源于大自然，你能在酒店里感受到中性的基色、鲜艳的色彩以及木本色。加上室内地毯以及墙壁上大片图案中生动的色彩搭配，与木制颜色形成对比。设计师打破了建筑外部刻板的风格，营造一种安静、温馨和多彩的内部环境。

　　酒店的家具大多现代时尚，整洁干净，色彩搭配鲜艳。舒适、耐用，当然还有款式是挑选家具的首要元素。委托人选择与当地艺术家合作来完成房间和公共区域的艺术装饰。

W BALI VILLAS AND
E-WOW SUITE 巴厘岛W酒店别墅和E-WOW套房

Designer/
AB Concept

Location/
Bali, Indonesia

W Retreat & Spa Bali – Seminyak asked AB Concept, an interior designer, to design a beautiful interior design called the W Bali Villas and E-WOW Suite Interiors. This villa is an ideal place where you can relax and enjoy every moment of the endearment. All rooms are good-looking and share an open feel of relaxing, a modernity which is impressive and showcasing a unique style of bringing the outside in. Inside the villa, the entire atmosphere seems to be the definition of the comfort and relaxation; the warm tones of blue, yellow, orange are inviting, as every little thing in this modern and cozy environment, which seems to be an image of the future.

巴厘岛塞米亚克W水疗度假酒店委托室内设计公司AB Concept 为巴厘岛W酒店别墅和E-WOW套房进行精美绝伦的室内设计。这栋别墅是一个理想的度假圣地，在这里你可以放松心情，尽情享受愉悦的每一分钟。每一个房间的设计都很好看，给人一种开阔、轻松的感觉。其独特的现代化设计风格，仿佛将外面的世界搬进了室内一般，让人过目不忘。整栋别墅的内部氛围让人觉得舒适而轻松自在。诱人的蓝、黄、橙等暖色调，犹如这个时髦又安逸的环境中的每个小物件一样，似乎都是未来的影像。

W HOTEL — SAN DIEGO

圣地亚哥W酒店

Interior Design/
Mr. Important Design

Location/
California, USA

Situated in San Diego city center, within walking distance of the San Diego Bay and close to the convention center, this hotel offers first-rate facilities, luxurious guestrooms and on-site dining options.

W Hotel – San Diego Rooftop Bar is perched on the tippy top of W Hotel – San Diego, while urban sophistication meets laid back style in the Living Room Lounge. At KELVIN, W Hotel – San Diego's new restaurant, contemporary world cuisine with a hint of Latin flair is provided.

W Hotel – San Diego features an outdoor pool and a fitness center. Guests can use the hotel's helpful whatever/whenever service for sporting events tickets, transport and dinner reservations. After a busy day, guests may relax with a massage in the Away Spa.

The guestrooms at W Hotel – San Diego boast pillow-top mattresses and Bliss spa products. In the evening, guests can refresh with a drink from the fully stocked minibar or snack from in the in-room Munchie box.

圣地亚哥W酒店位于圣地亚哥市中心，步行即可到达圣地亚哥湾并邻近会议中心。酒店提供一流的设施、豪华的客房和酒店内就餐服务。

屋顶酒吧设在热闹的酒店顶层，而都市的繁杂与闲散的生活方式在客厅式酒吧交汇。在酒店新开的餐厅KELVIN，顾客可以品尝到带有拉丁风味的当代世界美食。

酒店设有室外泳池和健身中心。顾客可以享受酒店内的随时/随需服务，包括预订赛事门票、提供便利交通以及预订餐位。一天繁忙的生活过后，顾客可以在Away温泉浴场享受按摩，放松身心。

每间客房配备的豪华席梦思床垫和Bliss品牌的水疗产品是酒店的一大特点。房间的迷你酒吧内酒水丰富，到了晚上，顾客可在此小酌以怡情或是品尝点心盒内的精致点心。

ONSEN PAPAWAQA

泰安观止

Location/
中国台湾

Onsen Papawaqa is a typical architecture of minimalism, whose outward appearance is architected by fair faced concrete and glasses. Iron wood railings are lined irregularly along the edge of the corridor, which bring rhythmic feelings to people when they walk or ride along the building. The facade of the building is a revolutionary design. The simple appearance seems barely visible. The main and the auxiliary buildings are constructed skillfully according to the terrain offering a pleasant surprise to visitors with their natural rough appearance.

The interior of the building showcases the typical feature of logs. The unique Taiwan incense cedar is largely used in the whole hotel, such as spas, basins, beds, floors, chairs, tea tables, TV stands and plaques in the rooms. Even the lobby counter is made of a whole piece of incense cedar. Unique solid wood materials, creative methods and the unique Pythoncidere essence together make up a romantic journey.

Onsen Papawaqa has 66 rooms of 28 types. Each room has an independent bathroom, spa and bed, with no conflict between the dry and wet areas. Except small rooms that only have a spa pool, all large and mid-sized rooms have both a cold pool and a spa pool. Room lights shine in warm tones and the luminance is deliberately reduced to 80% of the normal one in order to create a romantic atmosphere. The architect uses the bold grey series in the rooms to make a completely different taste and create mysterious and romantic surroundings.

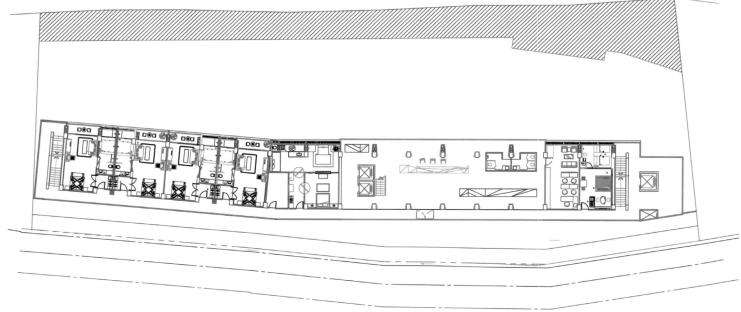

泰安观止是极简主义的经典建筑，清水模与玻璃架构了整个建筑外观。长廊的外延竖立着一根根不甚规则排列的铁木屏栅，在人骑车行走之余带来律动效果。建筑正面的设计脱离了传统思考模式，质朴的外观以"犹抱琵琶半遮面"的状态展现在世人面前。主副建筑巧妙地利用地形高低的视觉感，天然粗犷，让人惊奇连连。

建筑内部具有典型的原木特色。全馆大量采用台湾独有的肖楠原木，应用到房间的汤池、脸盆、床、地板、椅子、茶几柜、电视架和饰板中。就连大厅的接待台都是整棵肖楠原木制作而成。独特的实木材料、独创的工法、独特的芬多精香气，让人体验独特的浪漫之旅。

泰安观止共设66个房间，包含28种房型，每一房间内都设有各自独立的卫浴空间、泡汤空间及床铺空间，干湿分离。除小型房型只设有温泉池外，大中型房型分别设有冷池及温泉池。房间内的灯光均采用暖色系，还特意将亮度降低至正常亮度的80%，营造Motel的浪漫气氛。而房间则采用了大胆的灰色系列进行搭配，塑造完全不同的品位，更添神秘浪漫情境。

T HOTEL

T酒店

Design Firm/
Studio Marco Piva

Area/
20, 240m²

Location/
Cagliari, Sardinia, Italy

In T Hotel, the rooms have been designed to impress, tranquilize and pamper the guests. Thanks to the interior design by Marco Piva, harmony is combined with practicality to provide accommodation of marked impact and comfort.

The 207 rooms, Classic, Deluxe, Suite, T Junior and Executive Suite, arranged throughout the main hotel building and in the spectacular 15 floor tower with its breathtaking panoramic view.

Depending on the wing of location, rooms are themed on a prevailing color: vibrant orange, fiery red, relaxing green and tranquil blue.

T Hotel has two restaurants open to hotel guests and to the public: a main restaurant to seat 98 guests, a bar bistro to accommodate 50 people and a 270 setting banqueting hall.

T Bar for pleasurable moments of leisure enjoying a relaxing view over the winter garden surrounded by water effects.

T Conference Center includes a main hall with 300 seats and 6 meeting rooms, all with natural daylight, equipped with up-to-the-minute audio and video technology, simultaneous translation booths, control room and business center.

T酒店的房间设计旨在给客人留下深刻的印象,带来安定的心境并提供无微不至的服务。马瑞克·派沃独特的室内设计将和谐与实用性融为一体,酒店提供的膳宿给人带来强大的美学冲击,让人倍感舒适。

T酒店共有207个房间:标准间、豪华客房、套房、T标准套房和高级套房。房间遍布整个酒店主楼及壮观的15层塔楼内,在塔楼里可以俯瞰令人叹为观止的全景。

根据不同的房间定位,设计师采用了不同的流行色作为主题色调:活泼的橘色、热烈的红色、令人愉悦的绿色和宁静的蓝色。

T酒店内设两间餐厅,向酒店客人和公众开放:一间是可容纳98人的主餐厅,另一间是酒吧,可容纳50人,还可以布置成可容纳270人的宴会厅。

T酒吧为客人提供惬意的娱乐休闲时刻,在这里可以欣赏到处处布有水景的冬景花园。

T酒店会议中心设置了一间拥有300个座位的主厅和6间会议室。所有房间均为自然采光,内部配备了最新的音频和视频技术、同传箱、控制室和商务中心等设施。

TOWN@HOUSE STREET HOTEL Town@House 街区旅馆

Designer/
SIMONE MICHELI

Location/
Milan, Italy

Town@House Street Hotel is an ideal lodging of the era of twitter and facebook style virtual socialization. Town@House Street Hotel offers 4 suites only, one of them having 2 rooms, and therefore suitable for a family with children. Each color-themed suite features a different yet equally bright hue of green, red, yellow and orange. Each suite has its own entrance straight from the street. Instead of a key, you get a number along with your booking confirmation, which serves as an access code to your room. Someone perhaps might find it embarrassing, or quite the opposite – exciting, to sleep right behind a big shop window, with a mere curtain providing your privacy. The city sounds and noises penetrate the rooms and black and white photo wallpapers with Milan scenery emphasizes the urban environment surrounding you. Another unique aspect is that the entire trendy high-tech furnishing and fixtures are provided by sponsors (the list of them is displayed on the wall of each room). Lighting is adjustable to your taste, and each suite has a fully equipped kitchenette and a bathroom – like straight off the pages of design magazines.

在推特网和脸谱网这种虚拟社交方式盛行的时代，Town@House街区旅馆可谓是一个理想的住宿场所。Town@House街区旅馆仅提供四间套房，其中一间套房带有两间客房，适合有孩子的家庭入住。每间套房都以色彩为主题，各有特色。绿色、红色、黄色和橘色套房的色彩虽各不相同，色调却都鲜艳明快。每一间套房都有独立的临街入口。在进行房间确认时，你拿到的不是钥匙，而是一个数字，这个数字就是进入房间的密码。睡在商店的大橱窗对面却仅有一面窗帘来保护隐私，有些人可能会觉得颇为尴尬，或者正相反，有些人可能兴奋不已。城市中的声音和喧嚣透进房间，壁画上黑白的米兰风景照突显了周围的城市环境。旅馆的另一个独特之处在于，所有入时的高科技家具和设备都由赞助商提供（每一个房间的墙上都陈列着这些赞助商的名单）。灯光可根据个人的品位调节，每间套房的厨房和浴室都设备齐全，就像直接从设计杂志里搬来的一样。

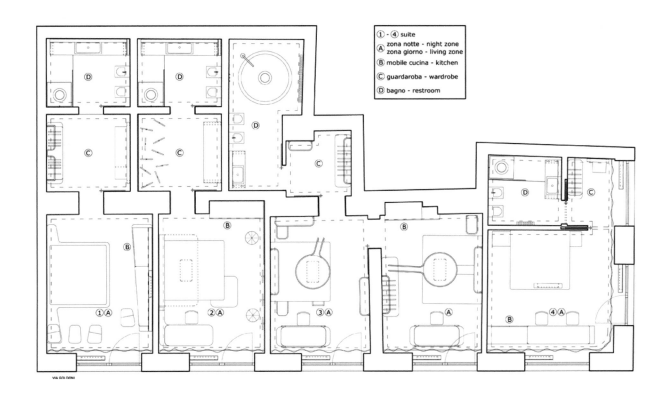

TOWN HOUSE STREET MILANO DUOMO 市政街米兰迎宾馆

Architectural Interior Planning/
SIMONE MICHELI

Location/
Via Santa Radegonda, 14 - Milan, Italy

Photographer/
Juergen Eheim

During Spring 2012 Town House Street Milano Duomo opens in Milan. This new hospitality space is situated in a former office building and is just a few steps from the Duomo Cathedral. These seven luxury 4 stars, new studio-rooms – "4 Studios" and "3 Double Rooms", located at Via Santa Radegonda, are designed by Arch Simone Micheli. The interior design is represented by the yellow color and the black and white pictures on the walls are portraits of the urban landscapes of this booming city.

All seven suites are dominated by yellow color: the excellent design view of Simone Micheli gives the entire project a unique atmosphere because of his work around geometrical furniture and object as well. The bed leans on a yellow color sofa which appears to float around it, bends and articulates on the same level forming two integrated bedside tables. A stainless steel tubular hosting the hangers is symmetric to a raised platform suitcase base. Chairs, sofas and tables generate a small area where one meets, relates and gets excited. The console can be used as a desk or a shelf: it depends on the guest's needs. All the suites are characterized by enormous mirrors braced on the wall, shaped as a macroscopic apostrophe run down and bent by a strong wind triggered by supernatural speed fitted with linear rear lights system and blue color LED. Thanks to the macro Photography of Maurizio Marcato which shows a breathtaking view of the city, the project has an exciting urban mood.

坐落于米兰的市政街米兰迎宾馆于2012年春天营业。这间新的迎宾场所位于前办公大楼内，与米兰大教堂只有几步之遥。这7间奢华的四星级新场所——4间工作室和3间双人房——建在米兰圣拉德公达街，由西蒙·米凯莉建筑事务所设计完成。室内设计以黄色为基调，墙上的黑白照片取自这座新兴城市的景观。

这7间套房全部以黄色为主色调：西蒙·米凯莉杰出的设计品位以及他对几何家具和物体的设计，给整个项目增添了独特的气息。床铺倚在一张黄色的沙发上。这张黄色沙发仿佛飘浮在它周围，沙发向下弯曲并在同一水平处相连，形成两个完整的床头柜。挂衣架用的不锈钢管与下面升起的放手提箱的平台对称。椅子、沙发和桌子形成了一个小区域，人们可以在那里会面、聊天、度过愉快的时光。房间内的小桌既可以当书桌也可以当架子使用：这取决于客人的需求。所有的套房都以镶挂在墙上的大镜子为特色，形状像一个巨大的撇号，好似被超自然速度引发的强风吹弯了一样，镜子装有线性背光系统和蓝色LED灯。Maurizio Marcato的放大摄影展示了令人惊叹的城市景观，为这个项目带来了令人激动的都市氛围。

BLOSSOM HILL — ZHOUZHUANG SEASONLAND 花间堂·周庄季香院

Design Firm/
Dariel Studio

Designer/
Thomas DARIEL

Location/
中国周庄

Area/
2,500m²

Blossom Hill – Zhouzhuang Seasonland Hotel is located in the watery region in the south of Yangtze River – Zhouzhuang, which is only 1.5 hours' drive from Shanghai and a perfect place for retirement and leisure. This boutique hotel was reconstructed on the base of three old buildings with the style of Ming and Qing dynasties. Before rebuilding, the three individual buildings were used as a museum, a tea house and a hotel respectively. And a part of them had been already abandoned. Dariel Studio renovated the old buildings with great caution and turned them into a boutique hotel with 20 guest rooms, at the same time hoped to retain the original space structure and the historical heritage of the buildings.

The old buildings were divided into three parts, the eastern part, the western part and the middle part, built separately, however, connected with each other as a whole, featuring different structures and characteristics. It took Dariel Studio nearly half a year to renovate and reconstruct the building, including smoothening the ground, reinforcing the main beams, repairing and rebuilding the doors and windows, and re-dividing the structure. According to the requirements of the client, the boutique hotel will be integrated into the beautiful scenery and the long history of Zhouzhuang, presenting the peacefulness and elegance of the old town which has never changed, to preserve and retain the local history and culture.

The inner design of Blossom Hill – Zhouzhuang Seasonland Hotel is a combination of tradition and modernity. In this typical 19th century buildings, Thomas DARIEL combines the characteristics of ancient China with the modern furniture and the local artifacts, enabling visitors to be immersed in the soul and character of the city. The building is revitalized under the modern visual effects and the elegant atmosphere.

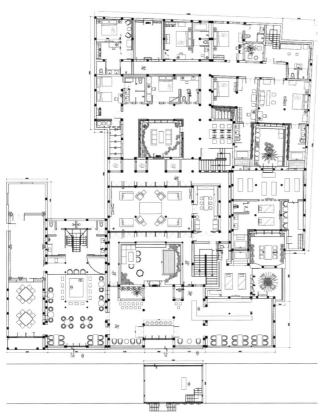

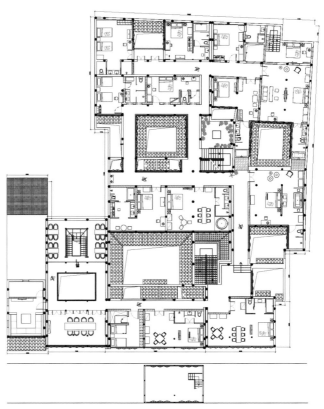

　　花间堂·周庄季香院酒店位于江南水乡——周庄，距上海仅1.5小时车程，是上海避世休闲的绝好去处。这个精品酒店项目由三幢明清风格的老建筑改造而成。改造前，这三幢独栋建筑分别被用做博物馆、茶室和客栈，并有一部分已经废弃。Dariel Studio非常谨慎地对这些古建筑进行了修复并将其合并改建成拥有20套客房的精品酒店，同时希望可以保留建筑最原始的空间结构及其历史传承。

　　昔日的建筑被分为东、西、中三宅，三宅独立而建，却又紧密相连，成为一个整体，格局迥异，各具特色。Dariel Studio用了近半年的时间对其进行修复改建，包括地面高低的统一、主梁的加固、门窗的修复和重建、结构的重新划分等。根据客户的要求，此精品酒店将与周庄如画的风景和历史相结合，展现古镇古往今来一直未变的恬静而优雅的特质，保留并延续当地的历史文化。

　　花间堂·周庄季香院酒店的室内设计融合了传统与现代。在这幢典型的19世纪建筑中，Thomas DARIEL将古老中国的特色与现代摩登家具交织在了一起，伴着当地特色的手工艺品，使游客沉浸于这座城市的灵魂与个性之中，并在现代的视觉效果与精致优雅的氛围下恢复活力。

117

GRAND HYATT SHENZHEN HOTEL 深圳君悦酒店

Design Firm/
Wilson Associates

Location/
Shenzhen, China

Photographer/
Chris Cypert Photography

The interior design approach for this hotel was entirely architecture-focused. Starting with an initial review of the building's massing and form, the interior architectural design team recreated the entire "crown" of the tower to resemble a stylized Chinese lantern with jagged angles enclosed with transparent glazing to allow unobstructed views outward toward the bustling city scene and luscious green mountain landscape.

The crown of the building on levels 33~38 showcases the hotel's sky lobby, lounge bar, and uniquely designed F&B outlets including the Show Kitchen, Belle-Vue and The Penthouse. The sky lobby space is the guest's first introduction to the dramatic double and triple volume voids, intentionally created to elicit a sense of awe and inspiration. By way of carefully crafted vantage points and strategically placed sightlines, views from one floor within the crown upwards or downwards to other floors reveal additional hotel spaces to discover and surprise.

Rooms and suites take on a rich contemporary design scheme, with oversized bathrooms accessible from either the corridor or bedroom. An earth-toned color scheme in the bedroom and living areas brings a sense of calm, while the floor-to-ceiling windows allow views of the city buzzing below. Total relaxation is found at the Shui Xiang Spa where holistic treatments are provided in the 11 treatment rooms that are situated over the podium roof garden. The double-panel system of opaque and transparent doors, the display kitchen successfully showcases the chefs as the centerpieces, working in a sleek black interior with brushed stainless steel worktops and equipment.

In the end, the design of this latest jewel of Shenzhen in positioned to shine for years to come.

　　酒店的内部设计方式完全以结构为中心。室内建筑设计团队首先对酒店的体量和外形进行了初步审视，重塑了塔楼的"冠"顶，使其类似中国的灯笼造型——带有锯齿状的角度，四周镶嵌着透明的玻璃。站在此处，外部城市的喧嚣景象和美丽的翠绿山景一览无余。

　　"冠"顶部分位于建筑的33至38层，展示了酒店的空中大堂、高级酒吧以及包括开放式厨房、悦景餐厅和阁楼的餐饮服务中心，设计独特。进入酒店，首先映入宾客眼帘的是2至3倍空间体量的空中大堂，令人产生敬畏感和灵感。通过精心设计独特视角，进行巧妙布局，让人们可以从冠顶任意一层向上或向下看向其他楼层，展现酒店更多额外的空间，让人们去发现、去感受惊喜。

　　客房及套房采取多样的现代设计方案，配有从走廊或卧室都可以进入的超大浴室。卧室和客厅选择大地色系，给人以安静感，而透过落地窗可以看到楼下熙攘的城市。宾客可以在水乡水疗中心尽情地放松。位于屋顶花园平台的11间理疗室内提供全套理疗服务。展示厨房的大门具有不透明和透明的双面板系统，成功地将厨师们作为展示的中心，他们在有拉丝不锈钢操作台和设备的整洁的黑色背景中工作。

　　最后，深圳这款最新设计的明珠，将会在未来的岁月中放射光芒。

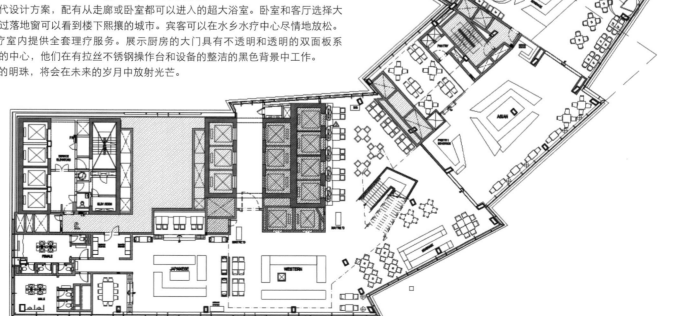

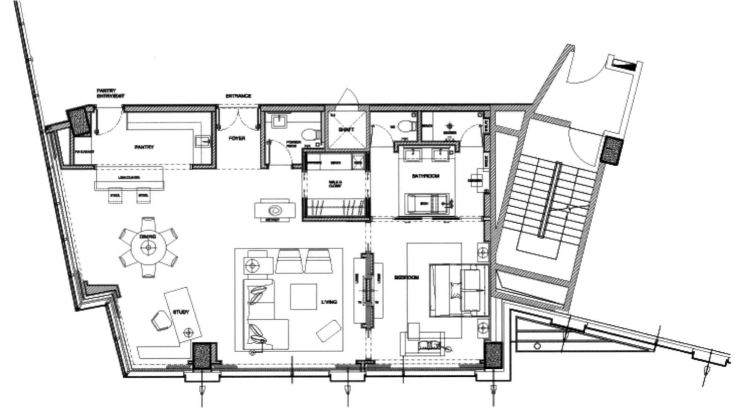

ONE & ONLY REETHI RAH

One & Only瑞提娜岛度假村

Designer/
Jean Michel Gathy

Location/
Maldives

One & Only Reethi Rah, designed by the world-wide known architect Jean Michel Gathy, was built along the 6km of private coastline with 12 pristine and fresh beaches. One & Only Reethi Rah has 130 luxurious villas, the largest in Maldives, some of which were built above the waters of the lagoon, including one-room villas or two-rooms. The villas were built around the islands, offering unparalleled views and unique privacy.

Reethi rah offers an excellent opportunity for each guest to explore these unique exclusive islands. If you need sheer relaxation and stretch, special massage and spa service is provided to let you feel the soft sea waves and the white sandy beach. A variety of entertainment facilities will make you active completely, enable you to make more new friends at the fascinating bar, or enjoy the delicious food to satisfy your taste at any one of the five romantic restaurants with different styles.

The hotel includes Reethi villas with exquisite decoration, Grand beach villas and water villas, all of which are elegant and spacious, of typical Maldives architecture. Inside the villa, well-equipped facilities provide guests with comfortable modern life. Under the clear moon lights guests may lie on the terrace to enjoy the sea sight, or to be waken up by the soft sound of waves, to feel the warmth of soft morning sunlight.

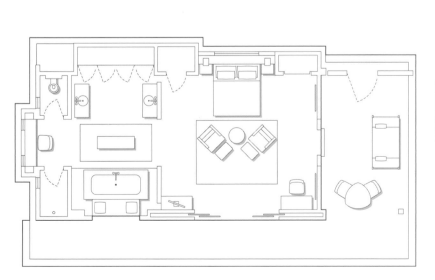
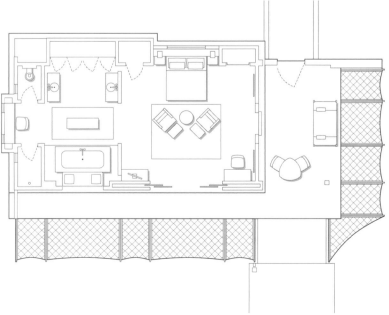

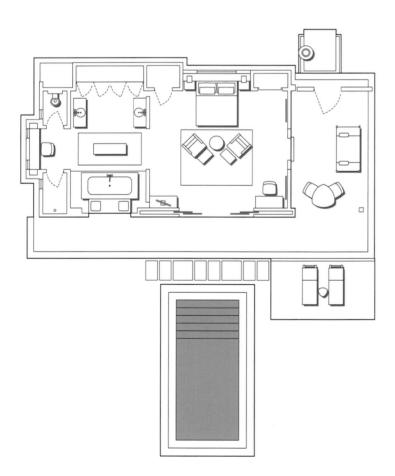

　　One & Only瑞提娜岛度假村由享誉国际的建筑师让·米歇尔·加蒂设计，沿6公里长的私人海岸线、共12个纯朴清新的海滩而建。度假村共设130间面积冠绝马尔代夫的奢华别墅，其中包括多间建于礁湖之上的一房及两房水上别墅。别墅环绕岛屿而建，为宾客提供无与伦比的迷人景观及私密空间。

　　瑞提娜海滩给每位客人提供了一个可以探寻这群独特岛屿的绝佳机会，因为这些岛都是瑞提娜海滩专有。如果你只是纯粹地希望放松和舒展，这里有专门的按摩舒压方式，感受轻柔的海浪和白色的沙滩。这里还有各种类型的娱乐设施让你彻底活跃起来，你大可在气氛迷人的酒吧里认识更多新朋友，或者在沙滩上5间风味各异、充满浪漫风情的任意一家餐厅享受美食，满足你的味蕾。

　　酒店拥有装饰精致的瑞提别墅、豪华别墅和水上别墅，雅致又宽阔，是典型的马尔代夫建筑。别墅内设施齐全，可以为客人提供舒适的现代生活。在皎洁的月光下，惬意地躺在露台上欣赏整个海景；或者被清晨轻柔的海浪声唤醒，感受朝阳柔和的光暖暖地洒在身上。

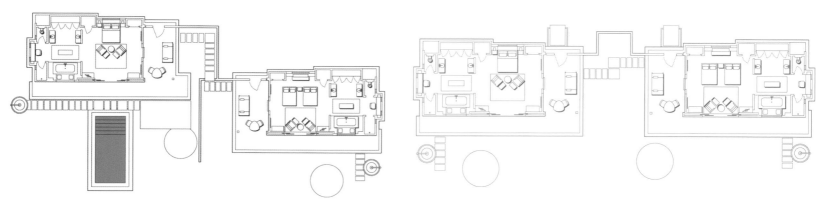

NEVAI HOTEL

那威酒店

Designer/

Yasmine mahmoudieh studio

Location/

Switzerland

Client/

King's Verbier

Area/

2,000m²

This retrofit inserts several new, varied programmatic areas and elements into a tight existing space. New layouts, new staircases, new connections between ground and first floors, new public bathrooms, and more rooms are part of a comprehensive re-thinking of the role and use of the typology of the alpine resort. The retrofit strives to relate to the surrounding mountain region through hinting at, never copying, the immediate environment or a typical alpine style. The design uses smooth lines, curved shapes, cool, calm tones, meticulous lighting, and abstracted images of nature in a contemporary and comfortable aesthetic, away from traditional expectations of "gemütlichkeit". This cosiness is achieved through a strong, contemporary visual language, which is repeated throughout the hotel: furniture, objects, seating areas, partitions between public and private.

The public areas of the hotel are designed with a conscious idea of openness, an openness that allows people from without to peek into the interior excitement and energy within. The entrance of the Nevai is dominated by the organic shape of the engraved corian reception desk, a clear sign and vessel of the symbols of the concept: ice, ice crystals, snow, and snow flakes. Special attention has been given to public seating areas: as a response to the low ceiling height in the existing structure, the intimate seating area facing the bar – dominated by a large gas fireplace – is sunken, protecting it against high-traffic zones. The French-inspired boudoir lounge, with its red seating, creates a cosy area that lends itself to intimate conversation or allows one to just relax quietly with a drink. The bar counter was inspired by a sled.

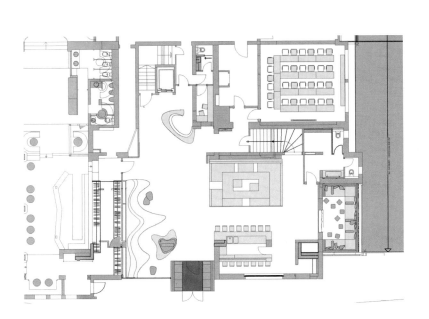
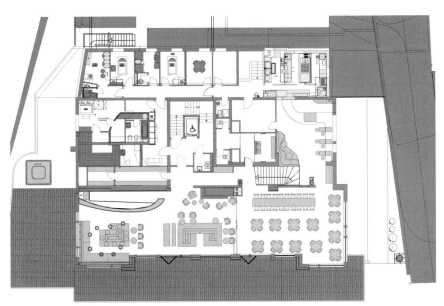

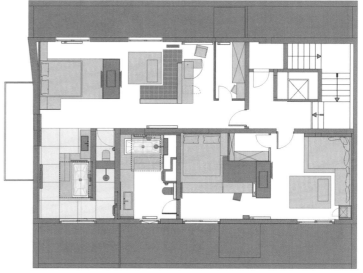

Suite

该酒店经过翻新成为现有的紧凑空间,嵌入了许多新式的不同活动区域和元素。全新的布局、崭新的楼梯、一层与二层的新式连接、新的公共浴室以及更多的房间设计都是针对阿尔卑斯度假村这一类型建筑的角色和用途进行全面考量的一部分。该酒店的翻新力图通过暗示邻近的环境或者典型的阿尔卑斯风格,与周围山区建立联系,而并非纯粹地复制这种环境效果。与传统的对"舒适环境"的期望不同,该设计采用了平滑的线条、曲线形状、清凉而平静的色调、细腻的光线以及运用当代舒适美学展示出的自然抽象画来表现。这种舒适感通过一种强烈的现代视觉语言展现出来。这种视觉语言在酒店中随处可见,包括家具、物品、休息间以及公共和私密场合的分隔区。

公共区域的设计体现了一种清晰的开放理念。这种开放性能让人从外部感受到内部的激情和活力。在那威酒店入口,最显眼的是雕花可丽耐大理石接待台。船形的接待台和清晰的符号凸显出概念的象征:冰、冰晶、雪和雪花。设计师还特别注意了公共休息区的设计:为了迎合现存结构中天花板较低的现状,面对酒吧(其中有个巨大的燃气壁炉)的私密的座位区下沉一些,以保证不受人多区域的干扰。设计灵感取自法国的会客室,以其红色的座位,营造出一种舒适的环境,适合亲密的交谈或独自小酌一杯,静静地在那里放松自己。吧台的设计则是受到了雪橇的启发。

HOTEL MISSONI
KUWAIT 科威特米索尼酒店

Designer/
Rosita Missoni, Mattheo Thun

Location/
Salmiya, Kuwait City, Kuwait

Set amidst extravagant seaside, Hotel Missoni Kuwait is an exclusive enclave, contrasting with both the vast sandscapes of Arabia and the customary opulence of Kuwait. Hotel Missoni is situated in the prime location of Symphony Center on Arabian Gulf Road in Salmiya which is the famous shopping district of the town.

The 169 uniquely designed rooms and suites by Rosita Missoni and the famous Italian designer Mattheo Thun embody the simplicity, elegance, and warmth of the hotel's graceful surrounds. All the rooms and suites have stunning views overlooking the Arabian Gulf Road. "Cucina" is an authentic Italian dining house for people to experience with highly sought after products to discover the first-hand flavors of the region. "Choco Cafe" is a unique delicatessen cafe concept focusing on chocolate and coffee beverages combined with sweet and savory fresh dishes, and "Luna" is a lively club restaurant on the 18th floor with stunning views overlooking the gulf. Laze around the outdoor pool or participate in such quintessential holiday pursuits as shopping, sightseeing and golf. For pure relaxation, embark on one of Six Senses' highly addictive, designer spa journeys.

与阿拉伯半岛广袤的沙滩和科威特那令人习以为常的财富相比，坐落于奢华海滨的科威特米索尼酒店拥有着独具一格的气质。该酒店位于塞尔米亚阿拉伯湾路交响乐中心的黄金地段，该地区是塞尔米亚著名的商业区。

由罗西塔·米索尼和意大利著名设计师马特奥·图恩共同设计的169间风格独特的客房与套房充分体现了优美酒店环境的简约、典雅和温馨。所有的客房与套房都有极好的视野俯瞰阿拉伯湾路的全景。"Cucina"是一家正宗意大利餐厅，提供的美食广受欢迎，让游客直接领略该地区的风味。"Choco餐厅"是一家独特的茶餐厅，主要提供巧克力和咖啡饮品，配以甜点和新鲜可口的菜肴。"Luna"是一家热闹的俱乐部餐厅，位于18层，拥有俯瞰海湾的绝佳位置。在露天泳池旁消磨时光，或购物、观光、打高尔夫，这些都是假日里最典型的活动。如果是为纯粹的放松，那么水疗中心会带你开始一个时髦的温泉之旅。

RADISSON BLU AQUA 雷迪森丽笙水绿酒店

Design Firm/

Graven Images

Location/

Chicago, USA

The Radisson Blu Aqua Hotel Chicago opened in November 2011, unveiling the first look at the design vision for the Radisson Blu brand in the United States. For the hotel interiors, internationally renowned design studio Graven Images presents a thought-provoking, contemporary style.

Graven Images created subtle and overt design references to the city of Chicago. Heavy steel work abounds, paying homage to the Windy City's reputation as the birthplace of the modern skyscraper. Chicago's famous brick buildings influenced the brick piers in the lobby, and the city lights viewed when driving into Chicago at night sparked the idea for the cast glass bricks in the "lobby street".

Graven Images wanted to move away from the typical hotel environment by making sure that the public spaces were socially inclusive – everything interlinks, with guests being able to move around easily within spaces that are the opposite of sterile and intimidating.

Design highlights include a 15.24m long fireplace in the lobby lounge and a 20 ton solid steel cantilevered staircase leading up to the restaurant, Filini – homage to Chicago's lifting bridges and its history of foundries and steel framed buildings. In order to create a fresh, contemporary restaurant interior that could compete in Chicago's burgeoning restaurant market, the designers manipulated the lobby space to create a mezzanine level for Filini, directly above the location that was chosen for its sister bar operation. A key move was to interface both directly with the external street as well as the internal streetscape and wider hotel lobby. This juxtaposition of the hotel lobby elements created an open flowing arrangement that is foreign to the language of typical hotel lobbies.

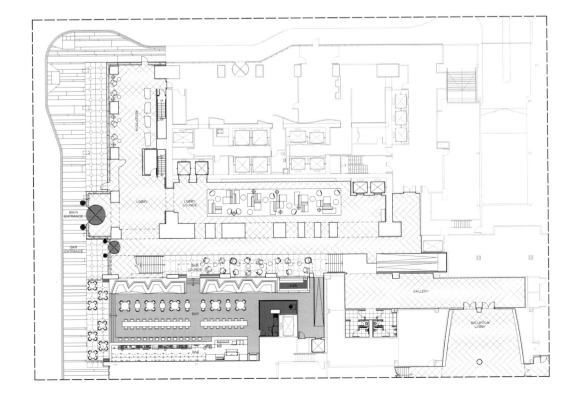

位于芝加哥的丽笙水绿酒店于2011年11月开业，首次揭开丽笙公司在美国的设计构想。享有国际盛名的Graven Images工作室通过对酒店内部的设计，呈现出一种发人深省的当代风格。

Graven Images工作室为芝加哥市创造了精妙而又显而易见的设计参考。大量的厚重钢结构，彰显出对风之城作为现代摩天大楼诞生地的敬意。芝加哥著名的砖块建筑影响到酒店大堂砖柱的建造，"大堂过道"浇注玻璃砖的设计灵感也是源于在夜晚开车进入芝加哥所看到的闪闪发光的夜灯。

Graven Images想要一改典型的酒店环境，使公共区域尽显社会包容性——所有的一切都相互联系，使客人们随意走动时不觉乏味，也不生畏惧。

酒店的设计特色包括大堂酒廊内15.24米长的壁炉和一个坚固的、重达20吨的钢制悬臂式楼梯，通向Filini餐厅，这一设计特色充分显示了对芝加哥升降桥、铸造厂历史和钢架建筑的敬意。为了创造一个新颖的当代餐厅内部环境，以便在芝加哥激烈的餐饮市场竞争中立足，设计师们利用大堂空间为Filini餐厅创造出一个夹层，恰好在其姐妹酒吧的上方。酒店设计的关键是将外部街道和内部过道景观与更为宽敞的酒店大堂直接联系在一起。这种设计不同于典型的酒店大堂设计，而是将酒店大堂元素并列在一起，营造出开放流动的布局。

NINGBO RITZ HOTEL

宁波丽兹酒店

Design Firm/

大勻国际设计事务所 / MoGA空间装饰设计

Designer/

陈亨寰、陈雯婧

Decoration Design/

祝戈

Location/

中国宁波

Area/

1,500m²

The furnishing of Ningbo Ritz Hotel is noble and elegant; the neutral color emphasizes its softy and warmness, expressing the unique ambience of this entertainment space. For the public areas, from the lobby, restaurant to coffee shop, the designers try to let the space showcase the same style. To strengthen the sequence of the original post in the building and separate the space by it. Chinese furniture for the space with hard lines and dark color has been widely used. Rooms are intended to create a sense of home; the overall tone of beige is soft and warm. Furniture, lighting and decorative details generate a warm living atmosphere.

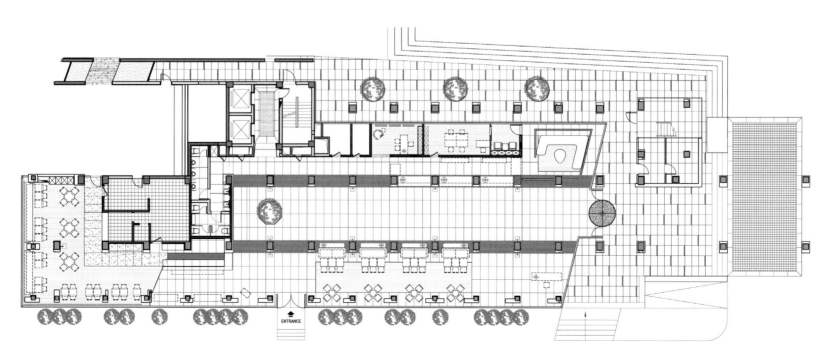

　　宁波丽兹酒店的室内装潢高贵典雅，淡雅的色彩平添柔美与温馨，尽显休闲空间的典雅意蕴。酒店的公共区域，从大堂、西餐厅到咖啡厅，流畅地展现出统一的设计风格。在强化建筑原有柱子的序列性的同时，将空间区隔开。酒店内的家具造型硬朗，以深色系打造，整体呈现中式风格。客房则着力营造出居家感，整体的米色调诠释，更显柔和、亲切。家具、灯饰和细节等装饰营造出温馨的氛围。

HOTEL NIKKO
SHANGHAI 上海日航饭店

Design Firm/
大观国际空间机构

Location/
中国上海

The designer employs three elements, music, season and art, to present the entire design concept and style of this hotel.

The first and second floors of the entrance lobby apply patio space, which places the primary imagery element on the central principal axis facing the entrance. When entering the hotel, visitors immediately see a Chinese-style colored-glaze vase; the stone columns on both sides are in the shapes of blossoming buds; while the dancing golden music notes in the staff pattern on the floor come from "Toccata and Fugue in F major", a famous work of Bach, the pipe-organ master. This is a combination of oriental and occidental cultures, as well as that of visual and sensory organs.

In the reception area, the cup-shaped independent reception counter which abandons the traditional one-piece design and tactfully stays away from the columns in the lobby, welcomes the guests in a relaxed and casual way. And the crystal behind the reception counter builds a magnificent atmosphere.

The background in the lobby bar applies modern techniques to manifest the conception of spring: Japanese cherry blossoms dominate the middle part; the artistic pendant lamps get inspirations from the scene in which the petals of cherry blossoms fall into water; on the floor is the carpet with water-wave totems. The arc-shaped Crystal Suite looks quite unique. Soft lines between planes and elevations, together with personalized cambered sofa and round bed, fully present what is luxury and fashion.

Topping the whole construction, the Executive Lounge on the 25th floor possesses a 270-degree view of the city. On the floor of the lounge is a carpet collaged by oriental and occidental ancient maps, while the furs on the columns tactfully convert ancient trade goods into modern space decoration materials. In the seating area, a large cambered book shelf extends along the gallery, on which there are various ornaments and books of different culture offered to people from different places. Above the seating area are some lamps which seem flying along the cambered windows as if you are in the business-class cabin flying to your destination.

本案设计师采用音乐、季节、艺术三种元素相结合的方式来展现酒店的整体设计理念和风格。

入口大堂一二层是挑空空间，设计将主要的意向元素安置在面对入口的中心主轴上。进入酒店，随即映入眼帘的是中式琉璃花瓶；两侧的石材柱体以花苞绽放的瞬间为造型；而地面五线谱图案那跳跃的金色音符，取自管风琴大师巴赫先生著名的《F大调托卡塔与赋格》乐谱。这是东西方文化的结合，也是视觉与感官的结合。

接待区内宛如杯形的独立接待台，摒弃了传统一体式的设计，也巧妙地避开了建筑上的柱体，以一种轻松休闲的方式接待客人。接待台后面的水晶石营造出华丽的氛围。

大堂吧的背景运用了现代手法表现春天的意境：中间部分以日式樱花为主体；吊顶艺术灯的灵感来自樱花花瓣落入水中的瞬间情境；地上铺有水纹图腾地毯。水晶套房的弧形设计别出心裁。平面与立面上柔美的线条，搭配极具个性的弧线形沙发及圆床，尽显奢华时尚。

25层的行政酒廊位于整个建筑的最顶部，拥有270度绝佳的都市景观。酒廊地面铺有东西方古地图拼贴完成的地毯，而柱体上的皮草则巧妙地将古代的商业交易转换成极具现代感的空间素材。进入座位区，巨大的弧形书廊沿着廊道延展，上面摆放着各种文化的书籍及饰品，供不同地域的人士阅读。将具有飞行意向的灯具设置在座位区上方，沿着弧形的窗户仿佛在商务舱内飞向目的地。

BEIJING SHOUZHOU HOTEL 北京寿州大饭店

Design Firm/
合肥许建国建筑室内装饰设计有限公司

Designer/
许建国

Co-designer/
陈涛、欧阳坤、程迎亚

Location/
中国北京

Area/
16,000m²

Photographer/
吴辉

Beijing Shouzhou Hotel is located in Beijing CCECC plaza in the West Railway Station with the building area of 16,000m². It has an advantageous geographical location and convenient transportation. Beijing Shouzhou Hotel is built as a high-quality international hotel with the function of catering, accommodation, entertainment and club. The hotel offers various kinds of deluxe rooms. The dining room comprising banquet halls and private rooms can serve 500 people at the same time. The design of the conference center can satisfy different types of conference. There is also a spa in the hotel that advocates healthy and green life.

The design of Beijing Shouzhou Hotel is based on Hui-style and the concept of post-modernism. The whole decoration is luxurious and elegant. While meeting the function of combining space, it creates a strong classical atmosphere. The whole building is simple, solid and magnificent, and showcases the compatible beauty of antiquity and nature.

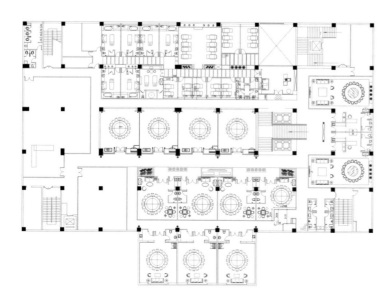
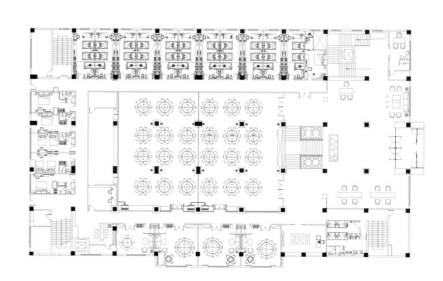

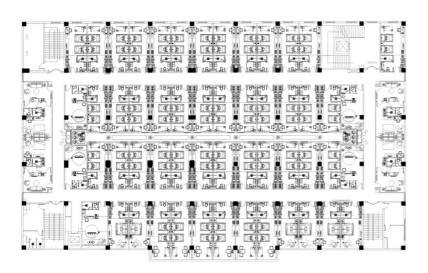
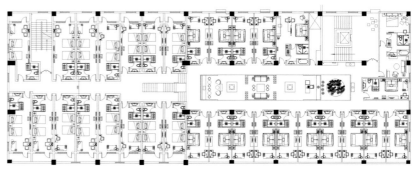

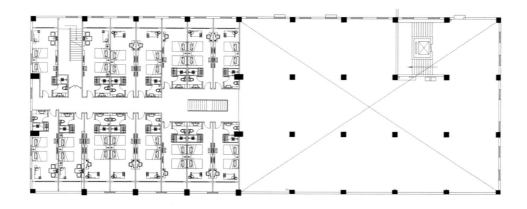

北京寿州大饭店位于北京西站中土大厦,建筑面积约16000平方米。其地理位置优越,交通便利。北京寿州大饭店以高品质涉外酒店要求建造,兼有餐饮、住宿、娱乐、会所等功能。饭店设有各类豪华客房,餐厅设有宴会大厅和包间,可一次性容纳500人同时用餐。会议中心的设计可满足不同类型的会议形式,并设有倡导健康、绿色的spa养生会所。

北京寿州大饭店以徽派及后现代理念设计,整体装潢豪华典雅。满足组合空间功能的同时营造出浓厚的古典氛围。整体设计古朴、沉稳、大气。古色古香,含蓄而自然,呈现出一种兼容并蓄的美。

LI MING HOTEL

黎明大酒店

Design Firm/
福州佐泽装饰工程有限公司

Designer/
郭宇宏

Location/
中国福建福州

Area/
6,300m²

Li Ming Hotel, covering a floor area of 6,300m², is located in the downtown of Fuzhou and a star-rated business hotel with an integration of accommodation and catering. Surrounded by various historical interests, the hotel enjoys the superior geographic position and elegant environment. The area is spacious, peaceful and beautiful, with ingenious pavilions, terraces and open halls appearing faintly in exotic plants, a brook murmuring beneath bridges, all coming into a coherent entity.

The hotel has deluxe suites, junior suites, banquet halls and dining rooms of various sizes. It also undertakes many large-scale activities, such as cocktail party, buffet dinner, fraternity and Chinese food banquet, offering international standard services. Guests will have access to all the recreational activities and physical exercises in the hotel. All the restaurant, coffee house, fishing club, piano bar, ecological dining room, sauna, nightclub, swimming pool as well as beauty salon are well-designed with modern artistic characteristics to meet the requirements of the guests.

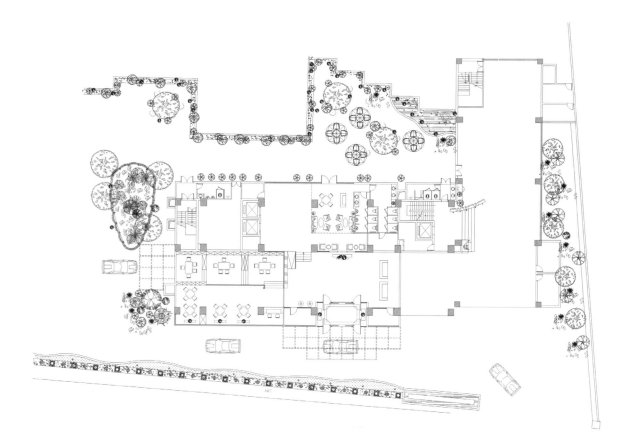

　　黎明大酒店地处福州市中心，占地面积6300平方米，是集住宿、餐饮于一体的星级商务酒店。酒店地理位置优越，环境优雅，周边名胜古迹荟萃。区内辽阔、宁静，自然风光优美，独具匠心的亭台楼阁隐现于奇花异草之中，小桥流水，浑然天成。

　　酒店内设各类客房，有豪华套间、套间、大小宴会厅及餐厅包间。可承办各种大型活动，如冷餐酒会、自助餐、联谊会以及中餐宴会，其各项服务都达到了国际标准。宾客无需出门便可尽享各种高雅的娱乐健身活动。各式充满现代艺术特色的餐厅、咖啡厅、钓鱼俱乐部、钢琴酒吧、生态餐舫、桑拿、夜总会、游泳池、美容会所，精心设计，尽应所需。

25HOURS HOTEL IN VIENNA 维也纳25小时酒店

Design Firm/

Dreimeta

Designer/

Britta Kleweken, Stefan Scheidecker

Location/

Vienna, Austria

Photographer/

Steve Herud, Berlin

A budget hotel shows personality: the 25hours hotel is a small idiosyncratic business card for Vienna which transports the city's surreal charm into its rooms. The circus theme acts like an entirely independent roof for the whole creative concept. The Circus motive runs like a thread through the interior design.

Under the family brand 25hours a local association to Vienna was to be created on the one hand, the basic principle of 25hours however had to function on the other. The house comes with all standards for a high-value offer in the low-budget sector and at the same time shows a completely new, a Viennese character in the 25hours hotel family.

The circus theme is a metaphor for Viennese cultural tradition – Vienna has been and still is the venue for sensations, shows, for the operetta-like, the shivaree with a folk-like impact. A circus symbolizes intoxication, merging into another world and gives the guests with a highly emotional sense of identification with dream images of childhood memories. As a basis for the design concept it was necessary to handle the theme on a maximum high-class level. The designers placed the nostalgia of the circus heydays during the early 20th century into contemporary and modern elements of today. They used a colorful and cheerful color concept. Authentic findings and wall paper with exclusive custom-made illustrations by the artist Olaf Hajek give each room its own impression. The bathroom lighting cites a typical artist's dressing room. Cage-like partitions structure the lobby. The entire house radiates in the lighthearted spirit of circus life.

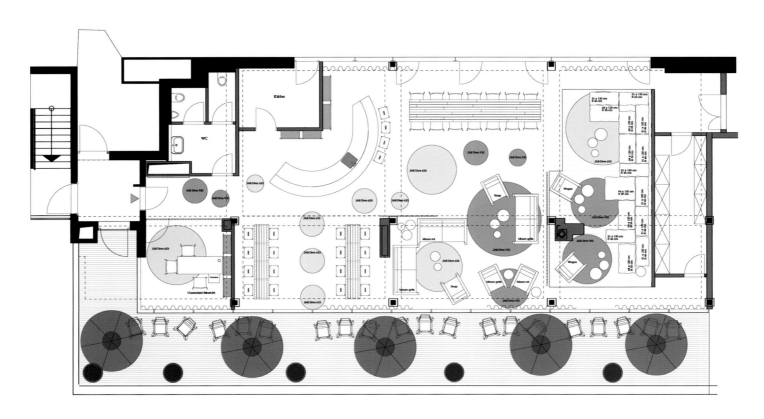

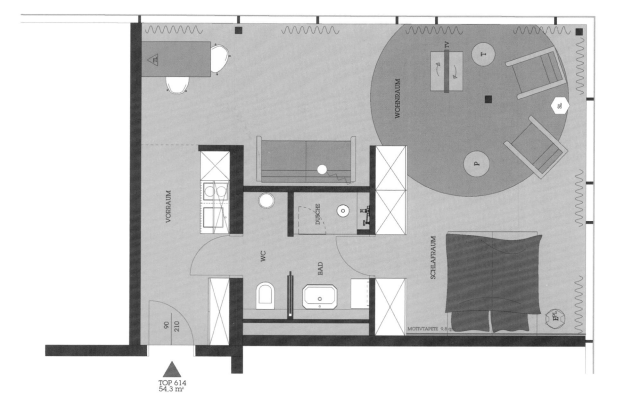

经济型酒店显示了其独特的个性：25小时酒店像是维也纳的一张小而独具特色的商业名片，将这座城市的超现实主义魅力传递到它的每一间客房。就其全部创意而言，马戏团的主题表现形式像整个创意概念中一个完全独立的屋顶。马戏团主题就像一根贯穿室内设计的主线。

一方面，作为家族品牌的25小时酒店，必然要与维也纳产生本土的关联；然而另一方面，酒店的基本原则也要发挥作用。这家家庭式酒店在提供高质量服务，保证经济实惠的同时，也展示了一幅全新的维也纳风情画卷。

马戏团的主题喻指维也纳的文化传统——维也纳一直是并仍旧是各种轰动事件、表演、小歌剧以及有民众影响力的喧闹城市。马戏象征着一种陶醉，融合到另一世界的陶醉，能够让客人们对童年记忆中那些梦幻景象产生强烈的情感认同。因此，将其作为设计理念的基础，在最高水平上处理关于主题的问题是有必要的。设计者将20世纪早期马戏全盛阶段的怀旧之情与现今的当代因素相融合，运用了一种生动、欢快的色彩概念。正宗的道具以及配有艺术家奥拉夫·哈耶克绘制的独特定制插图的墙纸，使每间客房风格迥然。浴室的照明设计参照了一位有代表性的艺术家的更衣室风格。大厅则设计成笼状的分区结构。整个酒店都洋溢出马戏表演般的轻松氛围。

CONSERVATORIUM HOTEL 音乐学院酒店

Design Firm/

Lissoni associates

Location/

the Netherlands

Photographer/

Amit Geron

The Conservatorium Hotel is comprised of 129 rooms and occupies eight floors with room categories in the property ranging from superior, deluxe, grand deluxe and suites of varying sizes. Paying homage to the building's original lofty ceilings and almost half of the rooms have been configured into stunning duplex layout with oversized functional windows and exposed structural beams. The rooms, all with natural daylight and double glazed windows, range in size from 30m^2 to a spacious 170m^2.

An integral pillar of the vision of The Conservatorium Hotel is the creation of restaurant and bar areas that will not just appeal to hotel guests but will become destinations in their own right for locals and international visitors, thereby imbuing the hotel with a wider responsibility in the local culture. The lap pool measuring 18m x 5m provides an oasis of calm and serenity within the comfort of the indulgent simplicity of Lissoni's design.

Outstanding furnishing from leading manufacturers such as Cassina, Living Divani and Vitra sit prominently in all communal spaces and are complemented by Piero Lissoni's own custom-made furniture and lighting while accent pieces such as vintage Asian rugs provide a sense of familiar comfort. According to Lissoni, "The sophisticated simplicity of the hotel was conceived with the hope of creating a personal atmosphere and ambience whereby guests and visitors alike instantly feel naturally at home."

音乐学院酒店共8层,拥有客房129间,根据性能不同划分为高级房、豪华房、超豪华房及面积不等的套房。为了迁就建筑中原有的高天花板,几乎一半以上的房间被设计成漂亮的双层布局,且都配置了超大实用的窗户和外露式结构房梁。这些装有双层玻璃窗的房间都有自然采光,面积从30平方米到宽敞的170平方米不等。

餐厅和酒吧区域的设计是音乐学院酒店的一个主要视觉景观。它们不仅吸引酒店的客人,而且也凭借本身的实力成为当地人和国际游客争相选择的目的地,从而使酒店在传播地方文化方面承担了更大的责任。18mx5m的小型健身泳池为人们提供了一个平心静气、安静悠闲的绿洲,这正符合里梭尼的极简设计风格中的舒适理念。

来自顶尖的制造商,如卡西纳、Living Divani和威达的精美家具摆放在公共空间内的显眼位置,皮埃尔·里梭尼也亲自定制了一些家具和灯具对这些家具进行补充,而一些有特色的物件如亚洲的老式地毯则给人一种熟悉的舒适感。正如里梭尼所说,"这种复杂中的简单特征是希望为酒店营造一种人性化的氛围,并希望这种氛围能使客人和参观者有宾至如归的感觉。"

KAMEHA GRAND BONN 波恩卡梅哈大酒店

Designer/
Marcel Wanders

Location/
Bonn, Germany

Marcel Wanders, the "rock star among designers" has created in each square of the Kameha Grand Bonn a potential "Lieblingsplatz" for every guest. With neo-baroque interior design with attention to the detail Wanders emphasizes the uniqueness of the Kameha Grand Bonn, creating a "place which is cool". A special highlight is the theme suites designed by Wanders that address the individual needs and desires of guests – while "Hero suite" with all the features and amenities of a modern office, the Fair Play suite with kicker and Wii, the Diva-suite with all the facilities for women, the Beethoven-suite with grand piano and an iPod, with the most beautiful concerts.

Marcel Wanders' design for the Kameha Grand Bonn is exceptional. In order to segment the large dimensions into smaller sections, Wanders created small, intimate areas that can be removed at any time when the large space is required. This was accomplished by using light, flowing materials that come down from the ceiling creating private "meeting islands" that exist within the midst of the vibrant life of the hotel. Exaggerations were also used to play with dimensions. This can be seen in the large chandeliers and the big golden flower vases; it adds a fun quality and a twinkle of the eye. The architect Karl-Heinz Schommer integrated the aspects sustainability, clear design with relevance to the neighboring river Rhine into the extraordinary shape of the building of the Kameha Grand Bonn. "With the building's eaves tapering down to the Rhine bank and the large terraces, we have created an unmistakably soft form for the building" says Karl-Heinz Schommer. Just like a Rhine wave, the Kameha Grand Bonn fits perfectly into the surrounding nature and Rhine landscape with unique view on the mountain range Siebengebirge.

马塞尔·万德斯被称为"设计师中的摇滚明星",在波恩卡梅哈大酒店的每个角落都为客人创造了其可能"最喜爱的位子"。酒店的室内设计为新巴洛克风格,且特别注重细节。万德斯强调波恩卡梅哈大酒店的独特性,将其设计得"十分时尚"。酒店的独特之处在于万德斯设计的主题套房,旨在满足客人的个人需求和愿望。例如,"男性套房"配备现代办公设备;游戏套房提供混合饮料和任天堂游戏机;女性套房提供全套的女性专用设施;贝多芬套房配有三角钢琴和苹果音乐播放器,提供最美的音乐会效果。

马塞尔·万德斯为波恩卡梅哈大酒店所做的设计可谓不同凡响。为了将大空间分割成小部分,万德斯创造了小型私密的区域,当需要大空间时,这些小区域可以随时被搬开。这样的设计通过使用从天花板垂下的轻质软性材料,在酒店充满生机的氛围中创造出私人的"聚会小岛"。同时,该设计在三维空间上也运用了夸张的方法。这一点在大型的枝形吊灯和巨大的金色花瓶上可以看出,不仅增添了乐趣,也让人眼前一亮。建筑师卡尔·海因茨·绍莫尔将建筑的持久性、简洁的设计等方面与相邻的莱茵河因素结合起来,设计出波恩卡梅哈大酒店独特的建筑造型。卡尔·海因茨·绍莫尔说:"建筑的屋檐朝着莱茵河河畔和大平台,呈坡状倾斜,这无疑为建筑创造出最柔性的外观。"波恩卡梅哈大酒店就像莱茵河里的波浪,完美地与莱茵河及周围的自然风景融为一体,从酒店望去,客人可以独享七岭山脉的美景。

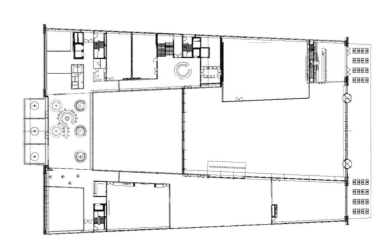

KEMPINSKI HOTEL

凯宾斯基酒店

Design Firm/
Wilson Associates
Location/
Dubai, United Arab Emirates
Photographer/
Michael Wilson

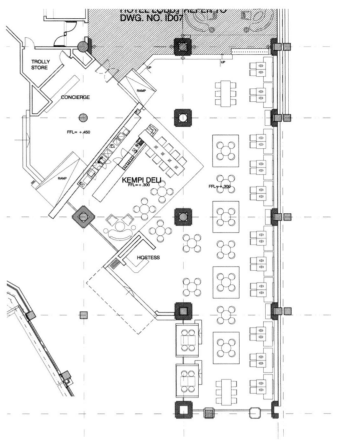

The hotel has a total of 400 rooms, comprising typical rooms, executive rooms, conference and leisure suites, penthouses, cabanas and ski chalets, the later being unique to Dubai. These rooms all face the indoor ski slope located in the snow dome. The main concept behind the design was to engage the travelers' five senses. This is translated in the form of highly textured surfaces, water features and backlit panels.

The Main Lobby is a feast for the visual senses. A "glowing" bridge that spans the void on the third story was added to draw attention upwards. Staggered panels of Indian Rosewood and paint, each of varying lengths, were added across the ceiling of the void. Concealed lighting was further added to "float" the panels away from the ceiling. Besides creating a fluid movement on the ceiling, the panels also turn vertically down the walls to form consoles and display ledges.

The design of the guest rooms is largely contemporary with Arabic influences added in artwork and artifacts. Colors are generally warm and inviting. The ski chalet suite design is a modern interpretation of a typical European chalet. Stacked ledge stone was used throughout the living, dining and bedroom areas to create texture and warmth while the combination of slate and marble was used in the bathrooms.

The Kempinski Hotel sets to define a new lifestyle destination and a central attraction in Dubai. The Hotel guarantees an unmatched "Alpine" experience.

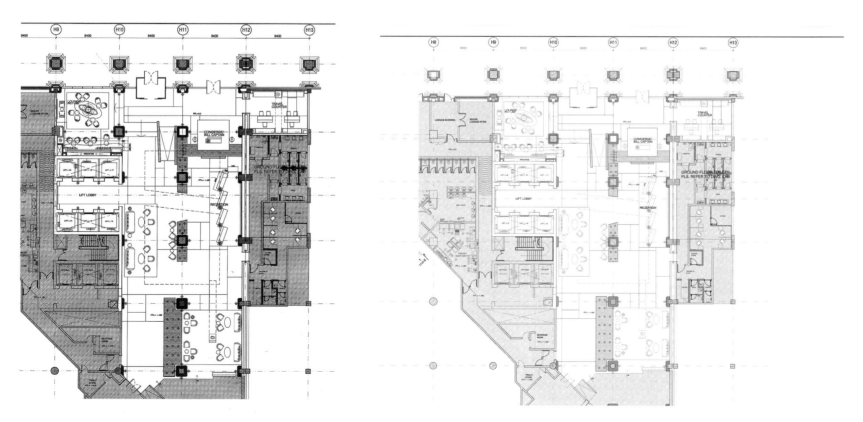

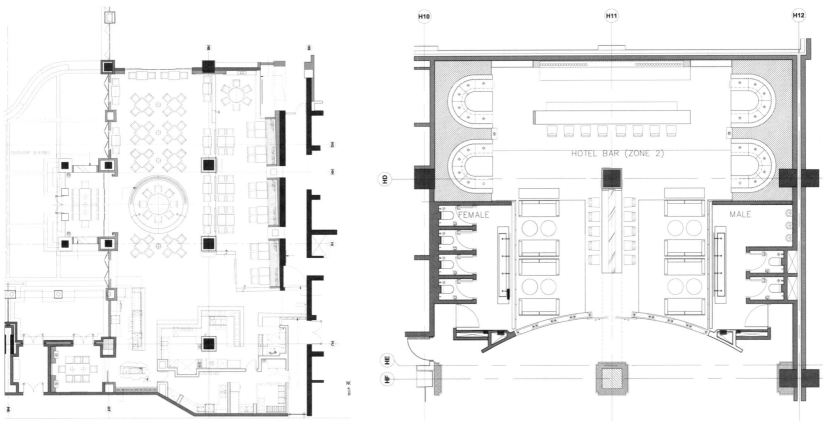

　　该酒店共有400间客房，包括标准客房、商务客房、会议室和休闲套房、阁楼、更衣室以及滑雪小屋。滑雪小屋在迪拜还是前所未有的。这些客房都朝向雪丘上的室内滑雪坡道。该设计的主要理念是为了充分调动旅客的五官感觉。这种理念体现在其高度质感的表面、水文特征和背光面板上。

　　大堂的设计仿佛一场视觉盛宴。在大堂上空第三层的位置架起了一座横跨空中的"发光"桥，将人们的视线引向上空。横跨大堂上空的天花板上，错列着长短不一的印度紫檀木板和漆画板。隐蔽照明的加入让这些面板与天花板之间产生若即若离的飘浮感。这些面板增强了天花板的动态感，同时也沿着墙面垂直摆放，形成支柱和壁架。

　　客房的设计融入了当代阿拉伯风格，具体表现在房间里的艺术品与手工艺术品上。客房的色调给人温暖的感觉，十分引人注目。滑雪小屋客房的设计充分展现了典型的欧洲小屋风格。起居室、餐饮室和卧室区域都使用了堆叠的礁石，以营造一种纹理质感和热情的氛围。浴室则采用了板岩和大理石的搭配。

　　凯宾斯基酒店的设计定义了一种全新的生活方式，成为了迪拜的吸引力中心。酒店能够带给人无与伦比的"阿尔卑斯山"的体验。

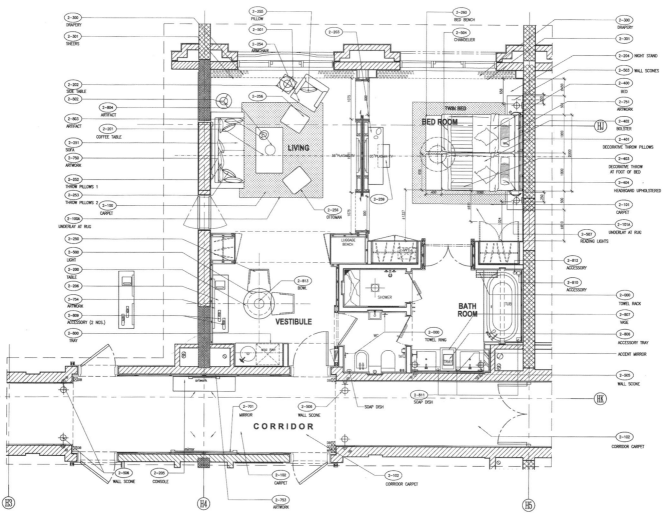

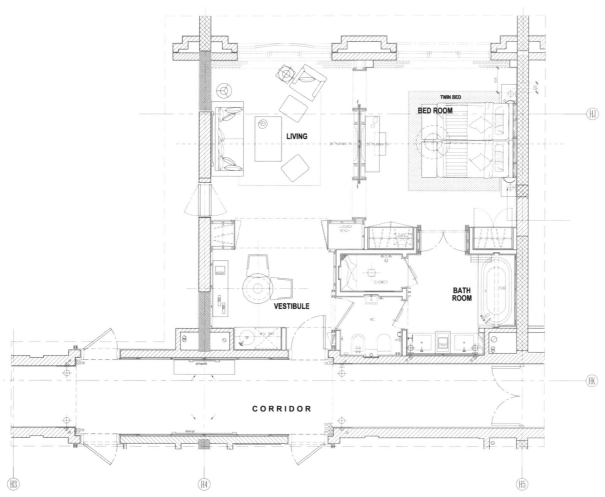

B.O.G. HOTEL

B.O.G.酒店

Designer/
Nini Andrade Silva

Location/
Bogota, Colombia

With a design inspired by Colombia's greatest natural treasures – gold and emeralds – the geometric B.O.G. Hotel, in the bustling north of Bogota, is a decadent base from which to explore this burgeoning city.
The stark yet luxuriously textured interior of the hotel complements the building's utilitarian exterior. All 55 rooms by award-winning designer Nini Andrade Silva focus on bronze, green, grey and beige combined with natural stone, bronze, mirrors, mosaics and tinted glass to create pared back and restful spaces. While the in-room décor is intentionally simple, every comfort is assured, including soundproofed windows, 500-thread linens and bathroom products and scents that have been exclusively designed for the hotel. After a day discovering the vibrant surroundings, the rooftop's heated swimming pool offers a welcome respite, with views over the city.

B.O.G.酒店的设计灵感源于哥伦比亚最知名的自然宝藏——黄金和绿宝石。这座几何体B.O.G.酒店位于波哥大繁华的北部地区，是探索这座生机勃勃的城市的起点。

朴实却奢华的内部结构与酒店建筑实用的外表相辅相成。55间客房都由备受赞誉的设计师妮妮·安德拉德·席尔瓦设计，青铜色、绿色、灰色和米色的主色调与天然石材、青铜制品、镜子、马赛克和彩绘玻璃结合，创造出一个简约而宁静的空间。虽然室内装饰朴实简单，却时刻给人面面俱到的舒适感，包括隔音窗、经纬密度为500的亚麻床单以及专门为酒店设计的浴室用品和清新剂。在热闹的酒店周围闲逛一天之后，屋顶的加热泳池是不错的休息之地，休息之余还可欣赏整个城市的风光。

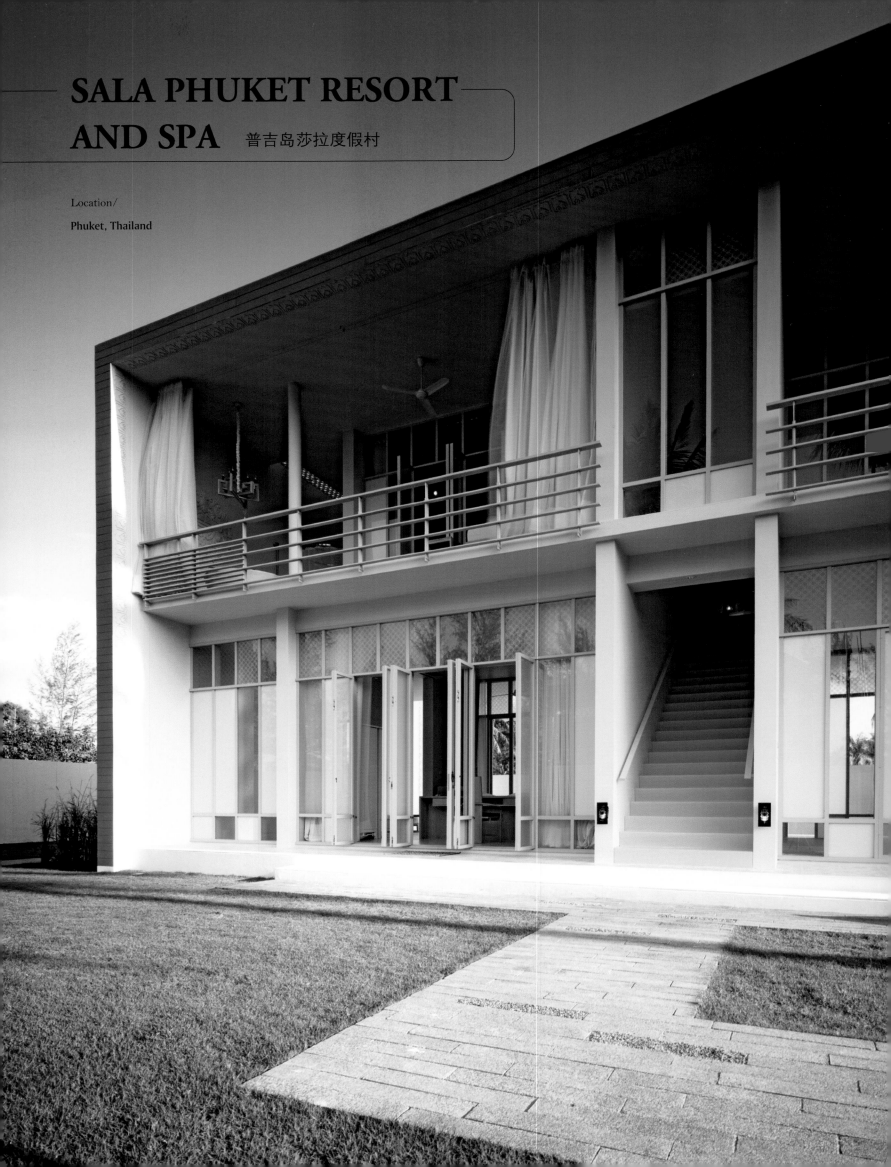

SALA PHUKET RESORT AND SPA 普吉岛莎拉度假村

Location/
Phuket, Thailand

Located on pristine Mai Khao beach on Phuket's northwest shoreline, Sala Phuket Resort and Spa is just 20 minutes from Phuket International Airport. Sala Phuket Resort and Spa is a stunning deluxe pool villa resort, featuring private swimming pools in 63 out of 79 rooms, villas, and suites. It combines rare historical Sino-Portuguese architecture with modern 5-star facilities. Sala Phuket luxury resort offers facilities including: 3 large beachfront swimming pools; World class international beachfront restaurant and bar with an air-conditioned dining option; Sala Spa; Gym; In-room Wi-Fi Internet connections and Business traveller services.

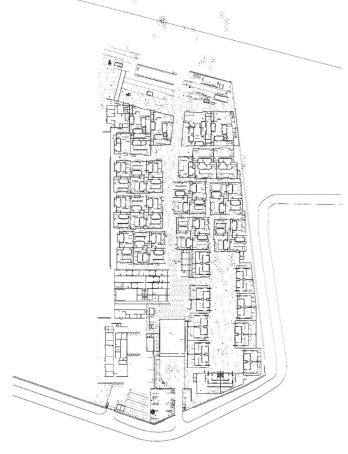

普吉岛莎拉度假村位于普吉岛西北海岸纯朴的迈考海滩,距普吉国际机场仅有20分钟的路程。普吉岛莎拉度假村是一个非常豪华的泳池别墅度假胜地,其独特之处在于79间客房、别墅和套房中有63间带有私人泳池。该度假村把罕见的中葡式历史建筑与现代化的五星级设施融合在一起。普吉岛莎拉豪华度假村所提供的设施包括:三座大型海滨游泳池、世界级国际海滨餐厅和可就餐的空调酒吧、莎拉温泉浴场、健身房、室内无线网络连接和商旅服务。

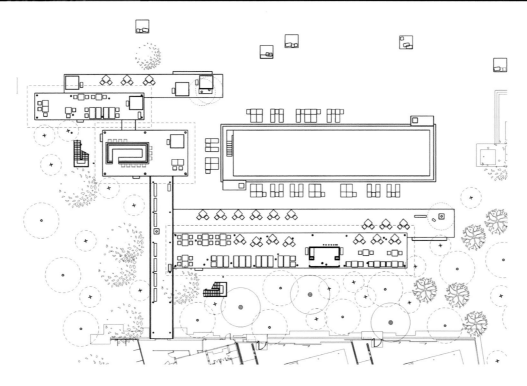

THE HOUSE HOTEL
BOSPHORUS 博斯普鲁斯House酒店

Design Firm/

Autoban

Designer/

Seyhan Ozdemir, Sefer Caglar

Location/

Ortaköy, Istanbul

Area/

1,000m²

Photographer/

Ali Bekman

This branch of The House Hotel is located at an attractive corner in a busy zone in Ortaköy, overlooking the breathtaking view of Bosphorus. Inspired by the historical Simon Kalfa building – which entitled its name from the Balyan Family who became famous by being architects of a dozen Ottoman palaces – the design concept was to create a comfortable, modern luxurious world reflecting classic traces of its era.

The whole atmosphere is fascinating with the latest custom-made designs through the use of Autoban's signature-materials as marble, brass, oak and walnut. Geometrical patterns seen in 3D cube shaped marqueterie on floors elongating on the carpets in corridor halls and guest room floors and ornamental decorative patterns enhancing the walls and other design details besides the circular forms used in furniture and lighting elements.

An elegant custom-designed metal framed door greets visitors at the entrance behind which the reception faces the waiting area. Hotel's restaurant & bar, gym, massage salon, conference room and lounge area are gathered on the first floor in accompany with guest rooms. Inspired by the historical buildings around the area, each of the 23 suites spreading over 4 normal floors and an attic floor has a special name in addition to room number. The furniture and the lighting fixtures are mainly sourced from the Autoban archive and adapted to the hotel concept. Custom-designed chandelier, fireplace and embedded bookshelves with concealed lighting generate warm and cozy atmosphere at the lounge and restaurant area where visitors can take a break and breathe in Istanbul.

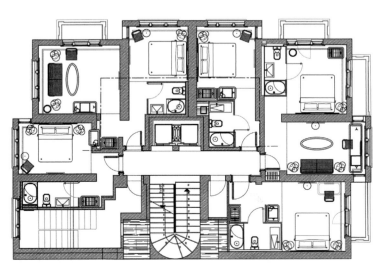
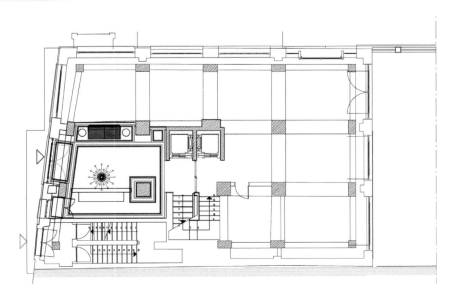

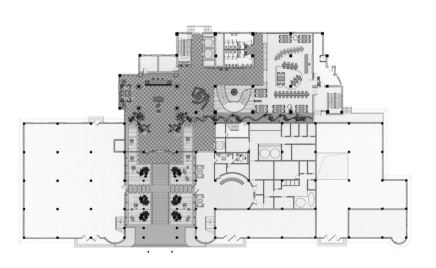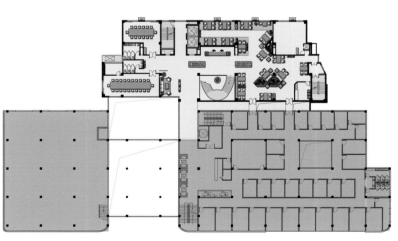

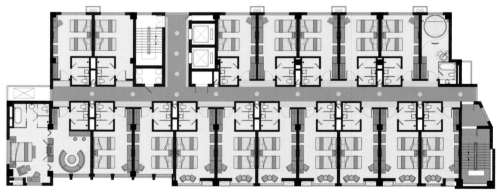
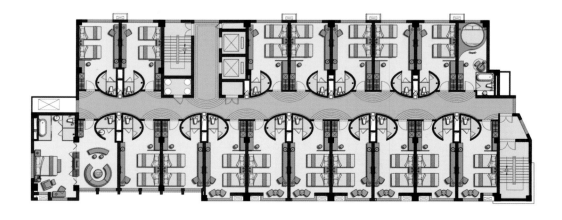

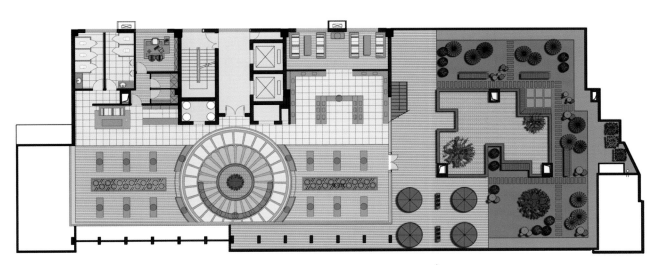

蝶尚非经验酒店是以人文艺术为主题、以星级酒店服务为标准、以海内外高端商务人群为目标客户的个性商务酒店。酒店共有110间客房，其最大的特点是将人文艺术主题活动和酒店功能融为一体，除酒店装修和服务处处体现与众不同的"人文艺术"特色，这里还可供黄山艺术群体举行沙龙聚会。设计师说："我们将会为顾客带来极具特色的消费体验，成功营造出与众不同的主题商务酒店。" 蝶尚非经验酒店的外立面、公共空间、春夏秋冬的主题客房都极力采用节能环保的LED光源，通过灯光营造出一年四季的场景，从感观上营造春暖、夏炎、秋高气爽以及冬凉的感觉，完全体现非经验的生物低碳建筑设计理念。

RAMADA PLAZA OPTICS VALLEY HOTEL WUHAN
武汉华美达光谷大酒店

Designer/
IEA 设计顾问

Location/
中国武汉

Adjacent to Optics Valley Square and Donghu scenic area, Ramada Plaza Optics Valley Hotel Wuhan is located in the core business circle of Optics Valley.

The guestrooms of the hotel are elegantly designed and fully equipped with modern facilities. The considerate service creates a high-quality leisure space for businessmen and travelers in their journey. With distinctive features, Chinese restaurant, western restaurant and Japanese restaurant, lobby lounge as well as the rotating restaurant on the top are elegant and splendid. Visitors can enjoy the delicious cuisines from all over the world and feel the exotic flavor.

The hotel provides banquet halls which can be as large as 700m² and various small and medium-sized multi-purpose conference rooms. As the rooms are well equipped with advanced facilities, such as simultaneous interpretation facility, various types of business conferences and banquets can be held here. There are series of fitness and entertainment facilities in the hotel, including gym, indoor swimming pool, salon, sauna, SPA, steam bath, KTV, chess and card room and auditorium. The perfect facilities will help to make the body and mind relax.

武汉华美达光谷大酒店地处光谷核心商圈,临近光谷广场和东湖风景区。

武汉华美达光谷大酒店客房设计高贵典雅,配备完善的现代化设备,体贴精心的服务,为商旅人士营造出高品质的旅途休闲空间。风格迥异的中、西、日餐厅、大堂酒廊及顶层旋转观景餐厅,氛围高雅,富丽堂皇,令您尽享天下美味,感受异域风情,给您带来完美的美食享受。

酒店拥有面积达700平方米的宴会厅及各式典雅的中小型多功能会议场所,功能齐全,拥有同声传译等先进设备,适宜举办各种商务会议及宴会活动。系列康体娱乐设施一应俱全,包括配备先进的健身房、室内恒温游泳池、美容美发、桑拿、SPA水疗、蒸汽浴、KTV、棋牌室、演艺厅等,设备精良,设施完善,让您舒展筋骨、尽涤身心。

SHANGRI-LA'S FAR EASTERN PLAZA HOTEL, TAINAN 香格里拉台南远东国际大饭店

Architect/

AB Concept

Location/

89 Section West, University Road, Tainan City 70146, Taiwan, China

Client/

Far Eastern Group

Operator/

Shangri-la Hotels & Resorts

As a grand arrival in a low-key town it was important for this award-winning five star hotel in Tainan to strike a balance between high sophistication and warm welcome. Though the proportions and the materials stay true to the glamour of the Shangri-La brand, certain techniques were used to bring the interiors down to a welcoming scale. The designers have avoided a sense of over indulgence by choosing a modest range of materials and fully exploring their uses and textures. Warmth from various local woods neutralizes the magnitude of the building, and connects it to the verdant nature of the town. Rough, interesting textures nurture an atmosphere of comfort as well as treat. However a sense of thrill is also encountered here, from the grand atrium and the imposing columns and water features of the lobby, to the sweeping ode made to the city by the hotel's 38th floor restaurant, reassuring guests that their regard for the Shangri-La brand is well placed.

对于坐落于小城市台南的大品牌香格里拉五星级酒店来说，在高档次和受欢迎程度之间取得平衡十分重要。虽然香格里拉台南远东国际大酒店的规模和建筑材料与香格里拉品牌的风格一致，但是设计师使用了某些技术进行了一些内部改变，使其达到令人舒适的程度。设计师选择了一系列普通的建筑材料，并充分挖掘了它们的用途和材质，从而避免了过度的奢华感。设计师为酒店选取了各种当地木材，其中散发出的暖意与翠绿的台南自然景观相呼应，中和了酒店带给人的恢弘感。木材粗糙有趣的纹理营造了一种舒适、快乐的氛围。然而，从宏伟的中庭和拥有壮观圆柱和水景的大厅，到赋予整个城市美誉的第38层餐厅，都带给人震撼感，香格里拉台南远东国际大饭店让客人有理由相信，选择香格里拉品牌是物超所值的。

ICELANDAIR HOTEL REYKJAVIK MARINA 雷克雅未克码头冰岛航空酒店

Architect/

Freyr Frostason, THG Architects

Location/

Myrargata 2, Reykjavik, Iceland

Icelandair Hotel Reykjavik Marina is located in the center of the Reykjavik harbour, adjacent to the main ship yard. The main concept is to connect to these elements surrounding the hotel, elevating and embracing the local condition for the global traveler. The graphics in the hotel show hints of the harbour history, the nature of the sea, the mapping of the bay and the culture of art and crafts displayed in various knots.

The rooms are in many sizes and shapes and each with its own character. Some with high ceilings and balconies with views over the harbour and mountains while others possess a city view. The first floor is the public area, creating various spaces and seating areas to accommodate for hotel guests as well as the local crowd. The first floor is open to the harbour with a pedestrian street connecting the hotel to the old city. There is a contrast in space creating, graphics and furniture creating a progressive environment where everyone can find their place.

雷克雅未克码头冰岛航空酒店坐落在雷克雅未克海港的中心，与造船主厂相邻。该设计的主要理念是将酒店周围的元素联系起来，提升并融合当地条件，为来自世界各地的游客提供更好的环境。陈列在酒店各处的平面艺术作品展示了海港的历史、海洋的自然景色、海湾的地形以及工艺品文化。

酒店的房间大小不一，造型各异，且各具风格特色。有的房间拥有高高的天花板，驻足阳台之上，可将海港和群山的美景尽收眼底；而在其他的房间则可以一览城市的风貌。酒店的一层是公共区域，为酒店住客和当地人提供各式各样的空间和座位。酒店一层面朝海港，通过一条人行小道与旧城区相连。酒店在空间、平面艺术作品和家具的设计上形成对比，营造出优异的环境，让每个人都能找到适合自己的空间。

UNA HOTEL

尤纳酒店

Designer/
Studio Marco Piva
Location/
Bologna, Italy

For the new hotel project, situated opposite Bologna railway station, the designer wanted to develop a particular concept that would enable him to "construct" a place dedicated to business tourism, located in the throbbing hub of one of the most active cities in Italy, still a pillar of thought and of Western history. Located in front of the central station, near the historic center and the trade-fair center, the new UNA Hotel Bologna, thanks to its strategic location is the ideal place both for business and leisure. UNA Hotel offers 93 rooms, 6 suites and 3 meeting rooms with design and comfortable facilities equipped with the latest technology, and fine restaurant services and relaxation in the bar with terrace.

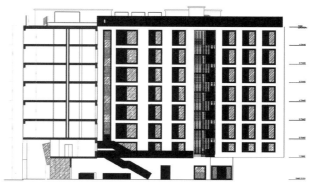

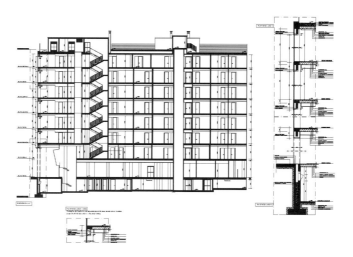

　　这个新的酒店项目位于博洛尼亚火车站对面,设计师希望将其建造成一个专门服务于商务旅游的酒店。酒店坐落在博洛尼亚的中心地带,该市是意大利最具活力的城市之一,也是思想和西方历史的核心所在地。

　　新的博洛尼亚尤纳酒店面对着博洛尼亚中心车站,靠近城市的历史中心和贸易展览中心,地处黄金地段,是商务和休闲活动的理想场所。酒店设有93个房间、6间套房,还有3间会议室,各个房间的设计构造和舒适的设备全部采用最新科技。带露台的酒吧可以提供良好的餐饮服务和休闲娱乐设施。

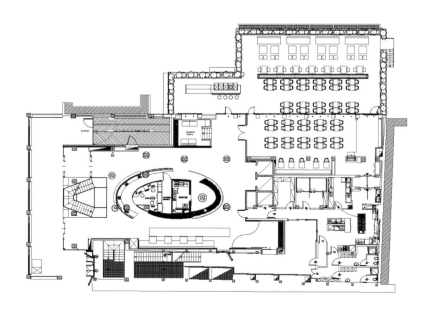

NANJING 21° THEME HOTEL 南京21°主题酒店

Design Firm/
陆定标空间设计

Designer/
陆定标

Location/
中国南京六合

Area/
1,100m²

Located in Liuhe District, Nanjing, Nanjing 21° Theme Hotel occupies a floor area of 1,100 m². Inspired by the image of cruises, the project incorporates the elements of cruises into its design which conveys the flavor of the sea with fashion and free romance. The magic purple lights and shadows, flowing lines, intensive color and leaping simple lines not only bring visual impact to guests, but also reflect the theme of the hotel. The cruise theme, endows the hotel with flowing space and changes. Besides, the strong sense of space and free romance arouse people's unlimited imagination. Walking into the hotel, guests feel that they are on a cruise at sea: seeing the clear blue sky above, feeling a breeze on their faces, standing in the hallway like standing on the deck… The designer satisfies people's infinite desire for sea by the scene design and decorative details. The roaring oncoming white cruise, splashing sprays and selective wallpapers make up a complete picture. The romantic flavor of the sea makes guests feel that they are on a cruise holiday…

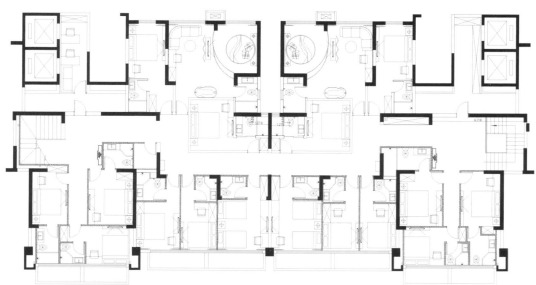

南京21°主题酒店建筑面积1100平方米，位于南京六合。本案以游轮为创意灵感，融入游轮元素进行设计。整个设计呈现时尚、自由浪漫的海洋气息。幻紫的光影、流畅的线条、强烈的色彩感与跳跃简洁的线条给客人带来视觉冲击，也体现了酒店的主题定位。酒店围绕游轮主题精心打造，赋予空间流动和变化，强烈的空间感和自由浪漫的气息给人无限遐想。走进酒店，犹如登上了一艘出海游轮：头上是清晰湛蓝的天，似乎还有阵阵海风迎面而来；站在过道，仿佛站在甲板之上……设计师将人们对海洋的无限向往，释放在酒店场景设计和装饰细节上。呼啸而来的白色游轮、恣意飞溅的浪花、壁纸的选用一气呵成，浪漫的海洋气息直抵人心，仿佛此刻正乘着游轮在海洋度假……

JIMBARAN HOTEL IN XIAMEN   厦门金巴兰酒店

Design Firm/
福州佐泽装饰工程有限公司

Designer/
郭宇宏、林金州、郑军、林振委

Location/
中国福建厦门

Area/
15,200m²

Jimbaran Hotel is located in Xiamen, Fujian Province, planned and constructed as a business and leisure boutique hotel based on high star standards, whose design is novel, unique and simple. The decoration is warm, elegant and fashionable with modern romantic glamour. The hotel provides more than 200 elegant rooms of different styles. Well-equipped conference rooms of different sizes will meet the needs for different class meetings. The exotic Bali western restaurant provides space for guests to taste the delicate European-style food; the ingenious coffee bar in the Aegean atrium with outdoor waterscape is also a beautiful scene for guests to enjoy. Comfortable and elegant atmosphere along with gentle and delicate service creates an "affectionate, leisure" place for guests to stay.

厦门金巴兰酒店位于福建省厦门市，酒店按高星级商务、休闲精品酒店标准设计和建造，设计新颖、别致、简约，装饰温馨、典雅、时尚，极具现代浪漫风情。酒店拥有200余间风格不一的精致客房；会议设施一应俱全，大小各异的会议厅可满足不同规格、档次的会议需求。充满异域风情的巴厘岛西餐厅为宾客提供了一个品味欧陆精致美食的空间；独具匠心的爱琴海中庭露天水景咖啡吧成为酒店里一道靓丽的风景线。清雅舒适的氛围、温婉细腻的服务，为宾客营造出"亲情、休闲"的停留环境。

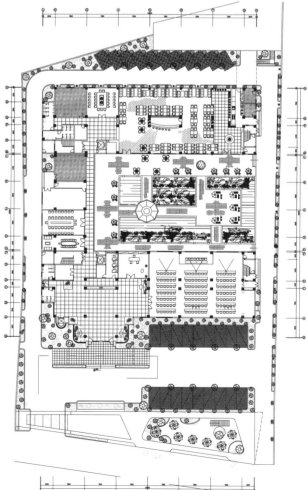
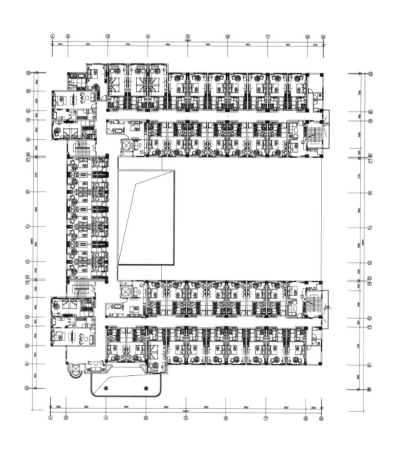

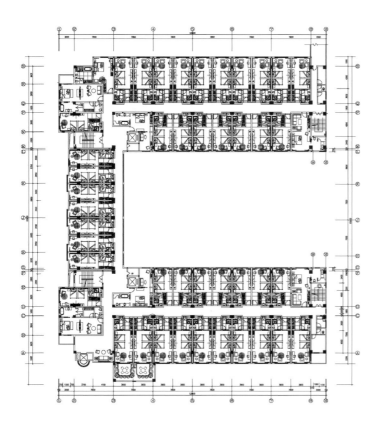
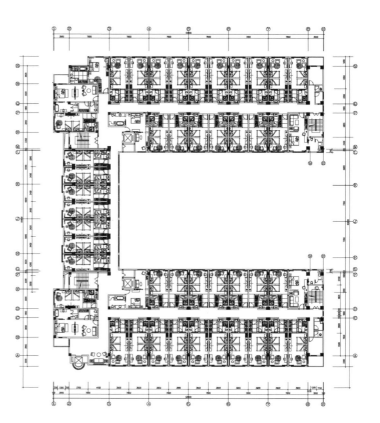

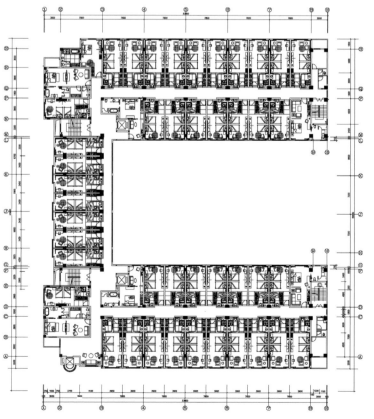
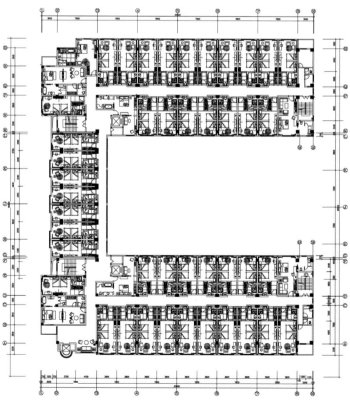

W HOTEL, LEICESTER SQUARE, LONDON 伦敦莱斯特广场W酒店

Architect/

Jestico + Whiles

Designer/

Concrete Architectural Associates

Location/

London, England

Area/

8,100m²

The interiors of W Hotel, Leicester Square, London have been designed by Concrete Architectural Associates, aims to offer guests a quintessential experience of London culture, by combining the formal and frivolous spheres of city life into one idiosyncratic experience. In a mix of traditional and contemporary materials, comfortable and cutting-edge design, Concrete's interiors represent the ever-evolving face of London glamour. Guests make their entrance underneath a disco ball cloud sculpture made of 280 disco balls, surrounded by black glass walls. On the left side of the entrance, Concrete's Spice Market restaurant mixes the ethnic vintage feel of Spice Market New York to the contemporary architecture of the new building. Once inside, visitors are led through the first floor reception atrium into Wyld, the stylish destination bar overlooking Leicester Square, providing an ideal location or back drop for film-premiere interviews.

伦敦莱斯特广场W酒店的室内设计由Concrete 建筑联合公司完成。酒店设计将城市生活的刻板和随意相融和，形成一种一体的特殊体验，使顾客能真正领略到伦敦文化的精髓。以传统材料和现代材料相结合，集舒适与前卫于一体，Concrete的室内设计代表了魅力城市伦敦的不断发展与演变。进入酒店，首先映入眼帘的是头顶上由280颗迪斯科球组成的云状雕塑，四周是黑色玻璃墙。酒店入口的左侧是东南亚餐厅，Concrete将纽约东南亚餐厅的民族复古感糅合进了这座酒店的当代建筑风格之中。顾客进入酒店后，穿过一层的接待中庭就是Wyld。Wyld是一间时尚酒吧，这里不仅可以远眺莱斯特广场，还是电影首映专访的理想地点或背景。

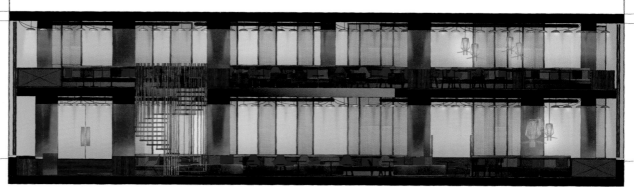

参与本书编写的人员：

徐宾宾、左昕、易蕾、罗玉婷、万浪斌、何志东、刘丹、王琪、余晶、苏丽萍、贾小帆、陈莉、杨玲、张永其、肖婧、张诗静、肖彬、金陈晨、吴瑕、凌波、许俊、李箫悦、李建生、李华、俞禄、王芸、杨亚龙、黄志恒、徐暑淑、邱为青、徐健、赵力、曹慧、丁迎春、李桥、刘涛、李新玲、袁志刚、张美祥、马勇、胡佳敏、刘露、曹紫青、简仕川、刘小宜